The business of space

The business of space

The next frontier of international competition

Louis Brennan

Associate Professor and Research Associate, Institute for International Integration Studies, Trinity College, Dublin, Ireland

&

Alessandra Vecchi

Postdoctoral Researcher, Institute for International Integration Studies, Trinity College, Dublin, Ireland

palgrave
macmillan

First published 2011 by
PALGRAVE MACMILLAN

Palgrave Macmillan in the UK is an imprint of Macmillan Publishers Limited, registered in
England, company number 785998, of Houndmills, Basingstoke, Hampshire RG21 6XS.

Palgrave Macmillan in the US is a division of St Martin's Press LLC,
175 Fifth Avenue, New York, NY 10010.

Palgrave Macmillan is the global academic imprint of the above companies and has
companies and representatives throughout the world.

Palgrave® and Macmillan® are registered trademarks in the United States, the United
Kingdom, Europe and other countries.

ISBN 978-1-349-31217-7   ISBN 978-0-230-30592-2 (eBook)
DOI 10.1007/978-0-230-30592-2

A catalogue record for this book is available from the British Library.

Library of Congress Cataloging-in-Publication Data
Brennan, Louis.
The business of space : the next frontier of international competition / Louis Brennan
and Alessandra Vecchi.
    p. cm.
Includes index.
Summary: "This book looks at the space industry from a business perspective, with a focus
on international competition. The space industry traces its origins to the middle of last
century as a government/military domain and the author now looks at the ongoing
evolution of space exploration and travel, and projects the future of the industry" —
Provided by publisher.

1. Aerospace industries.   2. Competition, International.   I. Vecchi, Alessandra.   II. Title.
HD9711.5.A2.B74 2011
338.4'76294—dc22   2011007009

10   9   8   7   6   5   4   3   2   1
20   19   18   17   16   15   14   13   12   11

CONTENTS

Figures

Tables

This book considers the business of space from the perspective of globalization. The space industry traces its origins to the middle of the twentieth century as an exclusively government/military domain involving the United States and the former Soviet Union. It has evolved to one that is increasingly commercialized and internationalized, encompassing a host of activities and countries. This book describes this ongoing evolution, assesses the major segments of the industry and evaluates the industry from a business perspective.

As the space industry is increasingly evolving from one dominated by governments and their military establishments to one that is undergoing rapid commercialization across a wide number of areas, there is a need for a business perspective on the industry. This book offers insights on this changing landscape that have relevance for the industry players, as well as policy-makers and decision-makers. Awareness of the industry is critical if public policy is to support it.

As the impact of climate change makes terrestrial survival for the human species increasingly problematic, the imperative to develop the means to evacuate planet Earth and sustain extra-terrestrial human existence becomes critical. Since the space industry is crucial in this respect, this book will provide a timely exposition around the emerging state of the industry.

The book comprises four chapters. Chapter 1 presents a conceptualization of the business of space from a globalization perspective. Chapter 2 considers the current state of the sector, while Chapter 3 deals with the main country players. The fouth and final chapter addresses the future by assessing the future potential of space applications, as well as the forces and key strategic issues faced by the industry.

ACKNOWLEDGMENTS

We express our appreciation to our International Business students for their invaluable assistance in sourcing pertinent data. We would also like to acknowledge the assistance of Ruslan Rakhmatullin in the production of the illustrative materials. Finally, we are greatly indebted to Palgrave Macmillan, in particular Eleanor Davey Corrigan and Keith Povey (with Elaine Towns) for their sustained and sustaining support and expert guidance in the writing of this book.

LOUIS BRENNAN
ALESSANDRA VECCHI

The authors and publishers are grateful to the following for permission to reproduce copyright material: The Organisation for Economic Co-operation and Development for tables from OECD publications *Space 2020: Exploring the Future of Space Applications* (http:/dx.doi.org/10.1787/9789264020344-en) and *The Space Economy at a Glance 2007* (http:/dx.doi.org/10.178 7/9789264040847-en); Taylor & Francis Publishers for figures from two articles by Maxim V. Tarasenko. Every effort has been made to contact all copyright-holders, but if any have been inadvertently overlooked the publishers will be pleased to make the necessary arrangement at the earliest opportunity.

Introduction and background: the global space industry

The conceptualization of the business of space: a globalization perspective

Since the early 1970s, debates have raged throughout the social sciences concerning the process of 'globalization' an essentially contested term whose meaning is as much a source of controversy today as it was in the 1980s, when systematic research first began on the topic. Contemporary globalization research encompasses an immensely broad range of themes, from the new international division of labor, changing forms of industrial organization and processes of urban–regional restructuring, to transformations in the nature of state power, civil society, citizenship, democracy, public spheres, nationalism, politico-cultural identities, localities and architectural forms, among many others. Yet despite this proliferation of globalization research, little theoretical consensus has been established in the social sciences concerning the interpretation of even the most rudimentary elements of the globalization process (for example, its historical developments, causal determinants and socio-political implications).

Nevertheless, within this whirlwind of conflicting perspectives, a remarkably broad range of studies of globalization have devoted detailed attention to the problematic of space, its social production, and its historical transformation. Major strands of contemporary globalization research have been permeated by geographical concepts (for example, 'space–time compression', 'space of flows', 'space of places', 'de-territorialization', 'glocalization' the 'global–local nexus', 'supra-territoriality', 'diasporas', 'translocalities' and 'scapes', among many other terms). Meanwhile, globalization researchers have begun to deploy a barrage of distinctively geographical prefixes (for example, 'sub-', 'supra-', 'trans-', 'meso' and 'inter-'), to describe various emergent social processes that appear to operate below, above, beyond or between entrenched geopolitical boundaries. In particular, in the social sciences, the recognition that social relations are becoming increasingly interconnected on a global scale necessarily problematizes the spatial parameters of those relations, and therefore also the geographical context in which they occur.

According to Friedman (2000), globalization is a new international system. It came together in the late 1980s and replaced the previous international system, the Cold War system, which had been in place since the end of the Second World War. Friedman defines globalization as the inexorable integration of markets, transportation systems and communication systems to a degree never witnessed before – in a way that is enabling corporations, countries and individuals to reach around the world farther, faster, deeper and more cheaply than ever before.

Several important features of this globalization system differ from those of the Cold War system. Friedman examined them in detail in his seminal book, *The Lexus and the Olive Tree*. The Cold War system was characterized by one overarching feature, which was *division*. That world was a divided-up, chopped-up place, and whether you were a country or a company, your threats and opportunities in the Cold War system tended to grow out of who you were divided from. Appropriately, this Cold War system was symbolized by the Berlin Wall and the Iron Curtain.

The globalization system is different. It also has one overarching feature and that is *integration*: the world has become an increasingly interwoven place. In the first decade of the twenty-first century, whether you are a company or a country, your threats and opportunities increasingly derive from those to whom you are connected. This globalization system is characterized by a single word – *web*: the World Wide Web. So in the broadest sense we have gone from an international system built around division and barriers to a system increasingly built around integration and webs. In the Cold War we reached for the hotline, which was a symbol that we were all divided but two people were in charge – the leaders of the United States and the Soviet Union. In the globalization system we reach for the internet, a symbol that we are all connected and nobody is quite in charge. While during the Cold War, division caused rivalry and intense competition, in the globalization era there is the acknowledgement that we shall all succeed only on the basis of collaboration. In this context, collaboration can assume many forms – from alliances to collaborative networks – and can involve a wide variety of actors.

Everyone in the world is affected, directly or indirectly, by this new system, but not everyone benefits from it, which is why the more it becomes diffused, the more it also produces a backlash by people who feel overwhelmed by it, homogenized by it, or unable to keep pace with its demands.

The other key difference between the Cold War system and the globalization system is how power is structured within them. The Cold War system was built primarily around nation-states. One acted on the world in that system through one's state. The Cold War was a drama of states confronting states, balancing states, and aligning with states. And, as a system, the Cold War was balanced at the center by two super-states, two super-powers: the United States and the Soviet Union.

The globalization system, in contrast, is built around three balances, which overlap and affect one another. The first is the traditional balance of power between nation-states, as there are now many 'super-powers'. In the globalization system, the United States is seen as the sole and dominant super-power and all other nations are subordinate to it to some degree. The shifting balance of power between the United States and other states, or simply between other states, still matters very much to maintain the stability of this system as it shifts from a uni-polar world dominated by the United States to a multi-polar world in which the United States is increasingly viewed as *primus inter pares*.

The second important power balance in the globalization system is between nation-states and global markets. These global markets are made up of millions of investors moving money around the world with the click of a computer mouse. Friedman calls them the Electronic Herd, and this herd gathers in key global financial centers – such as Wall Street, Hong Kong, London and Frankfurt – which Friedman calls the 'supermarkets'. The attitudes and actions of the Electronic Herd and the 'supermarkets' can have a significant impact on nation-states today, even to the point of triggering the downfall of governments and the consequent emergence of new, dominant private actors and, as was experienced during the late 2000s, the virtual collapse of the global economy.

The third balance we must pay attention to is the that between individuals and nation-states. Because globalization has brought down many of the walls that limited the movement and reach of people, and because it has simultaneously wired the world into networks, it gives more power to *individuals* to influence both markets and nation-states than at any other time in history. Whether by enabling people to use the internet to communicate instantly at almost no cost over vast distances, or by enabling them to use the World Wide Web to transfer money or obtain weapons designs that normally would have been controlled by states, or by enabling them to go into a hardware store and buy a US$500 global positioning device, connected to a satellite, that can direct a hijacked airplane –globalization can be an incredible force-multiplier for individuals. Thus individuals can increasingly act on the world stage directly, unmediated by a state. So, according to Friedman, we have today not only 'many super-powers', not only 'supermarkets', but also what Friedman calls 'super-empowered individuals' (see Table 1.1).

Table 1.1 summarizes the main arguments that allow Friedman to indicate the main differences between the Cold War system and the globalization era. The dynamics that globalization entails and as they have been described by Friedman are well represented if we look at the evolution of the space industry.

Table 1.1   Friedman's argument

The cold war system	Globalization
Division	Integration
Rivalry & competition	Collaboration
Two super-powers	Many super-powers
	Supermarkets
	Super-empowered Individuals

Source: Adapted from T. L. Friedman (1999), *The Lexus and the Olive Tree: Understanding Globalization* (New York: Farrar, Straus & Giroux).

The following section will review the evolution of the space industry by looking at its different stages – from the Cold War to globalization – and their distinctive dynamics in order to provide a new conceptualization of the business of space that incorporates the globalization perspective.

De-territorialization and re-territorialization

Since the early 1970s, debates have raged throughout the social sciences concerning the process of globalization – an essentially contested term whose meaning is as much a source of controversy in the 2000s as it was in the 1980s. Contemporary globalization research encompasses an immensely broad range of themes, from the new international division of labor, changing forms of industrial organization (Cerny 1995) and processes of urban– regional restructuring (Florida 1996; Storper 1995), to transformations in the nature of state power (Glisby and Holden 2005), civil society (Van Rooy 2004), citizenship (Castles and Davidson 2000), democracy (Goodhart 2001), public spheres (Bennett *et al.* 2004), nationalism (Hannerz and Featherstone 1990), politico-cultural identities (Brenner 1999; Ferguson 2005), localities (Amin 2002) and architectural forms (Kobrin 1997), among many others.. There is, however, a lack of consensus on the strength and likely impact of globalization. This conceptual vacuum is well described by Scholte (2000, p. 1), who affirms that:

> in spite of a deluge of publications on this subject, our analysis of Globalization tends to remain conceptually inexact,

empirically thin, historically and culturally illiterate, normatively shallow and politically naïve. Although Globalization is widely assumed to be crucially important, we generally have a scant idea of what, more precisely, it entails.

The contemporary era of globalization has been represented (Brenner 1999a) as the most recent historical expression of an ongoing dynamic of continual de-territorialization and re-territorialization that has underpinned the production of capitalist spatiality since the first industrial revolution of the early nineteenth century. On the one hand, capitalism is impelled to eliminate all geographical barriers to the accumulation process in pursuit of cheaper raw materials, fresh sources of labor, new markets for its products, and new investment opportunities. This expansionary, de-territorializing tendency within capitalism was clearly recognized by Karl Marx (Marx and Engels [1848] 2002), who famously described capital's globalizing nature as a drive to 'annihilate space by time' and analyzed the world market at once as its historical product and its geographical expression.

On the other hand, as David Harvey has argued, the resultant processes of 'space–time compression' must be viewed as one moment within a contradictory socio-spatial dialectic that continually moulds, differentiates, deconstructs and reworks capitalism's geographical landscape (Harvey 1991, p. 168). According to Harvey, it is only through the production of relatively fixed and immobile configurations of territorial organization, including urban built environments, industrial agglomerations, regional production complexes, large-scale transportation infrastructures, long-distance communications networks and state regulatory institutions that the capital circulation process can be continually accelerated temporally and expanded spatially. Each successive round of capitalist industrialization has therefore been premised upon socially produced geographical infrastructures that enable the accelerated circulation of capital through global space. In this sense, as Harvey notes, 'spatial organization is necessary to overcome space' (p. 168).

This theoretical insight enables Harvey to interpret the historical geography of capitalism as 'a restless formation and re-formation of geographical landscapes' (p. 168) in which configurations of capitalist territorial organization are continually created, destroyed and reconstituted as provisionally stabilized 'spatial fixes' for each successive regime of accumulation (Harvey 1991).

From this perspective, the business of space can be seen as a presupposition, medium and outcome of capitalism's globalizing developmental dynamic. Space is not merely a physical container within which capitalist development unfolds, but one of its constitutive key dimensions, continually constructed, de-constructed, and re-constructed through a historically specific, multi-scalar dialectic of de- territorialization and re-territorialization.

Building on this theorization, Brenner (1999b) understands globalization to be a double-edged process through which: (i) the movement of commodities, capital, money, people, images and information through geographical space is continually expanded and accelerated ('de-territorialization'); and (ii) relatively fixed and immobile socio-territorial infrastructures are produced, re-configured, re-differentiated and transformed to enable such expanded, accelerated movement ('re-territorialization'). Globalization therefore entails a dialectical interplay between the drive toward space–time compression under capitalism (the moment of de-territorialization) and the continual production of relatively fixed, provisionally stabilized configurations of territorial organization on multiple geographical scales (the moment of re-territorialization). The business of space and its geopolitical implications represent an emblematic example of this process by which de-territorialization and re-territorialization are taking place. Figure 1.1 illustrates these processes.

In this context, territorialization happens on planet Earth, where the geopolitical boundaries of individual countries are established; de-territorialization happens mainly through space exploration and re-territorialization takes place through individual countries establishing their presence either in orbit by the establishment of relatively fixed organizational forms (such as, for example, the establishment of a space station) or on other planets.

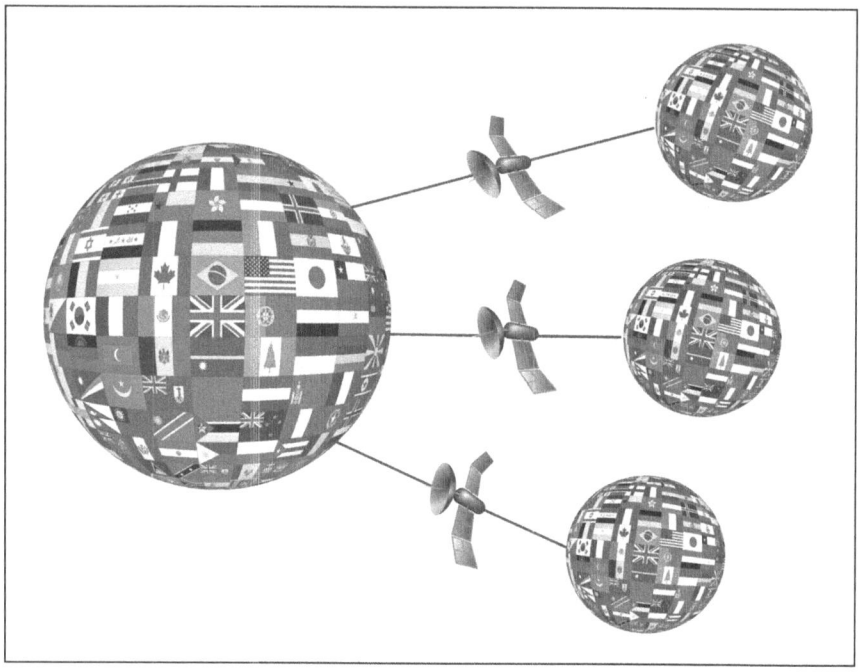

Figure 1.1 De-territorialization and re-territorialization in space

Origins of space

Human fascination with the world beyond the Earth's atmosphere pre-dates even the pioneering astronomers of ancient Greece. Great Stone Age structures such as Stonehenge in England are believed to have fulfilled astronomical (as well as religious) functions. Indeed, ever since humans first saw birds soar through the sky, they have wanted to fly. The ancient Greeks and Romans pictured many of their gods with winged feet, and imagined mythological winged animals.

The ancient fascination with the ocean of the skies is illustrated in the legend of Daedalus and Icarus, in which father and son escaped prison by attaching wings made of wax to their bodies. Unfortunately, Icarus flew too near the sun, the heat caused the wax to melt and he plummeted into the sea as a punishment for excessive daring. Chinese Legend tells us of the first attempt to propel a man into space made in the fourteenth century by a Chinese official, Wan Hu. He built a spacecraft with a chair, kites and forty-seven gunpowder-filled bamboo rockets. There was a loud explosion followed by smoke, and Wan Hu was never seen again (Burrows 1999).

Early developments

During the centuries when space travel was only a fantasy, researchers in the sciences of astronomy, chemistry, mathematics, meteorology and physics developed an understanding of the solar system, the stellar universe, the atmosphere of the earth, and the probable environment in space. In the seventh and sixth centuries BC, the Greek philosophers Thales and Pythagoras noted that the earth is a sphere. In the third century BC the astronomer Aristarchus of Samos asserted that the Earth moved around the sun. Hipparchus elaborated information about stars and the movements of the moon in the second century BC. In the second century AD Ptolemy of Alexandria placed the Earth at the center of the solar system (the Ptolemaic system).

Not until some 1,400 years later did the Polish astronomer Nicolaus Copernicus systematically explain that the planets, including the Earth, revolve around the sun. However, the scientific study of rockets, airplanes and satellites was inaugurated in the modern era by the impressive study of flight by Leonardo da Vinci, one of the most versatile geniuses of the Renaissance. His surviving notebooks contain over 35,000 words and some 150 drawings that illustrate his theories, and his sketches indicate advanced ideas regarding the parachute and helicopter, neither of which existed at the time. Later in the sixteenth century the observations of the Danish astronomer Tycho Brahe greatly influenced the laws of planetary motion set out by Johannes Kepler. Galileo Galilei, Edmund Halley, Sir William Herschel and Sir James Jeans were other astronomers who made contributions pertinent to astronautics. The scientific breakthrough in designing spacecraft came in the sixteenth and seventeenth centuries. Physicists and mathematicians helped to lay the foundations of astronautics, with the German mathematician Kepler (1571–1630), calculating the equations for orbiting planets and satellites, and Isaac Newton (1643–1727) formulated the laws of universal gravitation and motion, which confirmed that planets follow Kepler's equations. Newton's laws of motion established the basic principles governing the propulsion and orbital motion of modern spacecraft. In 1654, the German physicist Otto von Guericke proved that a vacuum could be maintained, refuting the old theory that nature 'abhors' a vacuum. In 1696, Robert Anderson, an Englishman, published a two-part thesis on how to make rocket moulds, prepare the propellants, and perform the calculations.

The nineteenth century brought about advanced research. The essential equations for rocketry were devised by a Russian school-teacher, Konstantin Tsiolkovsky (1857–1935). Tsiolkovsky concluded that space travel was a possibility and determined that liquid oxygen and hydrogen fuel rockets would be needed, and that these rockets would be built in stages. His prediction came true sixty-five years later, when the Saturn V rocket facilitated the first landing of human beings on the moon. Tsiolkovsky also stated that the speed and range of a rocket were limited by the exhaust velocity of escaping gases (NASA 2009). In America, Robert Goddard (1882–1945) produced the world's first liquid-fuelled rocket launch, in 1926. Goddard, 'the American father of modern rocketry', invented rocket technology and forged the designs that were used by Germans during the Second World WarI. After a confirmation of their military might and strategic importance during the war, it was Hermann Oberth (1894–1989) who convinced the world, in his highly influential and inter-nationally acclaimed book *The Rocket into Interplanetary Space*, that the rocket industry was something to take seriously. He was the only one of the three rocketry pioneers who lived to see men travel through space and land on the moon. Indeed, Oberth and a team of scientists, directed by Wernher von Braun (1912–77) developed and launched the German V2 rocket, the first rocket capable of reaching space. At the end of the Second World War, von Braun settled in the USA, where he played a crucial role in convincing the federal government to pursue a landing of humans on the moon, and guided US efforts to success (NASA 2009).

Literature has also greatly supported human interest in outer space. Greek satirist Lucian (second century AD) wrote about an imaginary voyage to the moon. Lucian was followed by French satirist Voltaire (*Micromegas* – the travels of inhabitants of Sirius and Saturn); French writer Jules Verne (*From the Earth to the Moon*); British novelist H. G. Wells (*The First Man in The Moon*; *War of the Worlds*); Sir Arthur C. Clarke (*A Space Odyssey*); Isaac Asimov (*Nightfall*), and many others. Despite the scientific foundations laid down in earlier ages, however, space observation and travel did not become possible until the advances of the twentieth century provided the actual means of rocket propulsion, guidance and control for space vehicles. The vivid interest in outer space and rapid technological progress that took place in the twentieth century turned dreams into reality. Expendable rockets provided the means for launching artificial satellites, as well as manned spacecraft. In the middle of the twentieth century, space became the new frontier of international rivalry.

The Soviet Union versus the United States

From the brief historical outlook presented so far, it can be seen that the efforts to develop and launch rockets or spacecraft into space gradually evolved from a personal fascination into more strategic attempts initiated and financed by states to serve their own potential military purposes. This was the first visible shift from 'territorialization' to 'de-territorialization' – before embarking on any space activity, geopolitical rivalry between states stemmed from asserting national borders on the Earth ('territorialization'), but the advent of space exploration took geopolitical rivalry to another level ('de-territorialization'). Figure 1.2 illustrates the most significant events that shaped the space industry over time.

The interstate rivalry reached its climax during the Cold War, when the two super-powers, the USSR and the USA, vied with each other by using their space achievements to communicate their might and prestige. The Gromov Flight Research Institute was founded in 1941 in the USSR, but the National Aeronautics and Space Administration (NASA) was not created until 1 October 1958 – 17 years later (BBC 2009). The goal of space exploration and the conquering of space leveraged tensions between the USSR and the USA to another level. On 4 October 1957, the USSR stunned the world by launching the first artificial

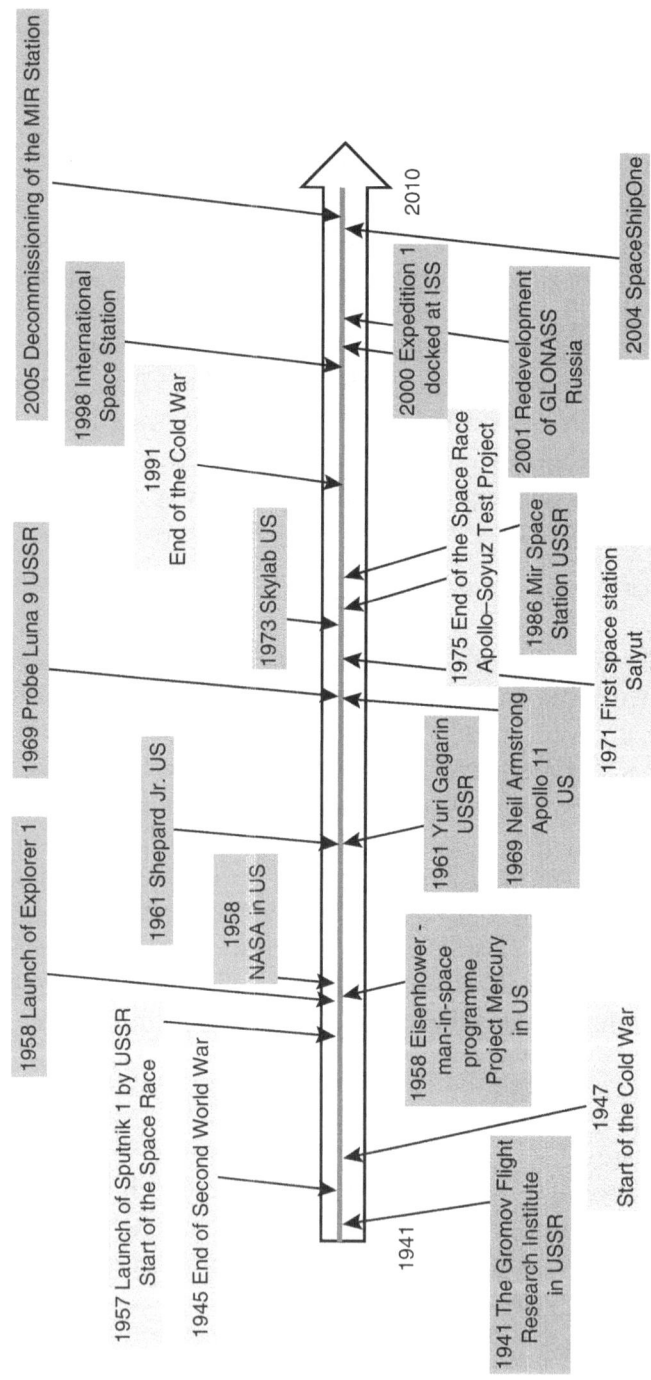

Figure 1.2   The evolution of the space industry

satellite, Sputnik 1. In the context of the Cold War, that launch catalyzed a whole series of events, ranging from the founding of NASA one year later to efforts at improving education in the USA, and the space race with its host of geopolitical implications (Dick 2007).

American president Dwight Eisenhower knew that militarizing space would only accelerate the nuclear arms race. Consequently, on 1 October 1958 he brought together research laboratories that were developing rocket technology and put them under civilian control (NASA 2009). In 1958, the USA initiated its first man-in-space program, the so-called 'Project Mercury'. Known today as the space race for control of the heavens, this battle produced some of the greatest technologies and outer-space heroes. The USA responded to the USSR's launch of Sputnik 1with the successful launch of Explorer 1 in 1958. Both of these satellites were intended for purely scientific purposes, but the space race was undoubtedly not just a quest for knowledge and scientific facts but also a manifestation of the Cold War and the desire to demonstrate US superiority over the Soviets. In the civilian sphere it was essentially a crusade for scientific prowess and proficiency, and to explore outer space with both satellites and humans, with the ultimate aim of landing on the moon. In a military capacity it was an extension of the arms race and the art of espionage, an investigation into rocket and artillery capabilities that had been carried out by German scientists long before the war propelled it to the fore. Since the space race was an integral part of the rivalry between the Soviet Union and the USA, both countries allocated large resources to developing the space industry. With the launch of Sputnik 1, the fear that it stirred, along with the realization that America did not have technological superiority in the field, quickly led to the formation of NASA by the USA. More importantly, great public interest had been aroused into what would soon develop into a 'star wars' era (Jones 2004).

Once these milestones had been achieved, the goal then became manned flight. The Soviet Union launched the first man into space – the cosmonaut Yuri Gagarin – on 12 April 1961. Three weeks later, the United States levelled the playing field and launched its first astronaut – Alan Shepard Jr. – into space.

Shepard was the first to manually control his spacecraft and he gave renewed hope to Americans within the Mercury project. Fifteen days after Commander Shepard's flight succeeded, President Kennedy issued the challenge to land a man on the moon by the end of the decade. He memorably said in his speech during a special session of Congress in 1961:

> I believe this nation should commit itself to achieving the goal, before this decade is out, of landing a man on the Moon and returning him safely to Earth. No single space project in this period will be more impressive to mankind, or more important in the long-range exploration of space; and none will be so difficult or expensive to accomplish. (Kennedy Project 2009)

This passage from John Kennedy's speech makes clear the importance he attached to the propaganda aspects of the US space effort (Jones 2004). According to Fisk (2008), the transformation of American society continued with President John Kennedy's remarkable pledge in 1961 committing the U.S. to place a man on the Moon and return him safely to Earth before the decade was out. Perhaps the most revealing statements of Kennedy's intentions appeared later in the speech when he said, 'a Moon landing would demand sacrifice, discipline, and organization: the nation could no longer afford work stoppages, inflated costs, wasteful interagency rivalries, or high turnover of key personnel'(Kennedy Project 2009). He also stated that 'every scientist, every engineer, every technician, contractor and civil servant must give his personal pledge that this nation will move forward, with the full speed of freedom, in the exciting adventure of space' (Kennedy Project 2009).

This endeavor was named the Apollo program. Kennedy viewed the Apollo program as an event that would transform the nation, and according to Fisk (2008) it did. At the peak of the Apollo program, NASA consumed 4 percent of the federal budget; some 400,000 Americans worked on Apollo; most being employed by some 20,000 American industrial firms of all sizes. From Apollo, and all the other aspects of space that developed concurrently, the US vastly improved its technical workforce, and its sense of what technology can accomplish (Fisk 2008).

After completing the first two objectives of sending a satellite and later a human being into space, the race rapidly became one to get to the moon. This was achieved by the Soviet probe Luna 9, but the first manned mission to the moon was on 21 July 1969, when Neil Armstrong became the first human to set foot on the moon's surface as part of the Apollo 11 mission. His words on stepping out on to the lunar surface have since become iconic – 'That's one small step for man, one giant leap for mankind' (Kennedy Project 2009). The Soviets were the first to orbit the Earth, to put both an animal and a human being into space and to reach the moon, but because they failed to walk on it, the 'race' is viewed by many to have been an American victory and the Soviet Union subsequently gave up the fight. The human landings on the moon were a set of signal historical events that had an impact on society, but it is not clear what the impact was or how long-lasting it was (Jones 2004). Even as the landings were taking place, Congress decided to terminate the program early, as the public had lost interest and funding priorities had changed (Dick 2007).

The USSR achieved a series of successful lunar orbiters starting in 1966; three lunar sample returns in 1970, 1972 and 1976; and two Lunokhod rovers in 1970 and 1973. In 1971 the Soviets launched the world's first space station, Salyut 1. The Americans followed with Skylab in 1973, but it fell to Earth five years later, killing a cow in Australia.

The conquest of the heavens was the primary political challenge during the Cold War period, but the space race ended in 1975 with the historic Apollo–Soyuz Test Project, during which an American Apollo spacecraft docked with a Soviet Soyuz spacecraft above the Earth for the first time (providing valuable information on the synchronization of American and Soviet space technology, which would prove useful in the future Shuttle–Mir Program). This began a new era of cooperative space ventures. Two of the space race's original heroes were present on that mission. Legendary Mercury astronaut Deke Slayton, and the first man to walk in space, Alexei Leonov, shook hands on 17 July 1975, symbolizing an era of cooperation and the end of space animosity up to that point between the two great powers. Both countries had poured huge amounts of money into their space programs, because many of the political and public opinion battles were being fought over superiority in space. Initially, space exploration was stigmatized by the nationalistic ambitions that fueled international space races.

Many events took place throughout the 1980s that contributed to the development of the space industry. In April 1981, Robert L. Crippen and John W. Young took part in the first mission in NASA's space shuttle program aboard the space shuttle Columbia. On 18 June 1983, Sally Ride become the first American woman in space with the launch of shuttle mission STS-7. On 30 August 1983, Guion S. Bluford, Jr. became the first black man in space, aboard the space shuttle Challenger. The journey into space has not been without its tragedies, however. On 28 January 1986, the Challenger space shuttle exploded 73 seconds after take-off with the loss of the seven crew members aboard.

In 1986 the Soviet Union launched the first module of the Mir space station, which was manned continuously by cosmonauts until it was decommissioned in 2005. The latest station (2010) is the International Space Station (ISS). This collaborative effort between NASA and RKA ('Roskosmos', the Russian Federal Agency) started in 1998 and ended in November 2000. Since that time the station has had a rotating permanent resident crew and has hosted five space tourists. Exploration too has continued since the end of the space race. For example, the European Space Agency's (ESA) Aurora Programme intends to send a human mission to Mars no later than 2030, and Russia has announced plans to build a permanently manned moon base.

However, with the end of the Cold War era, space exploration budgets in both US and Russia have shrunk dramatically (Jones 2004).

From space race to the business of space

The launch of the Mir space station in the 1980s played a major role in the transformation of the space industry. This was not only a significant landmark in the processes of 'de-territorialization' and 're-territorialization', but it was also a transformation characterized by a shift from fierce rivalry to global collaboration, as seen today in many areas such as the Galileo Positioning System project. (The Galileo Positioning System is a European project in which countries such as China, Israel, Ukraine, India, Morocco, Saudi Arabia and South Korea have all played a part.) Figure 1.3 depicts the main events of the space industry in relation to the processes of territorialization, de-territorialization and re-territorialization described so far.

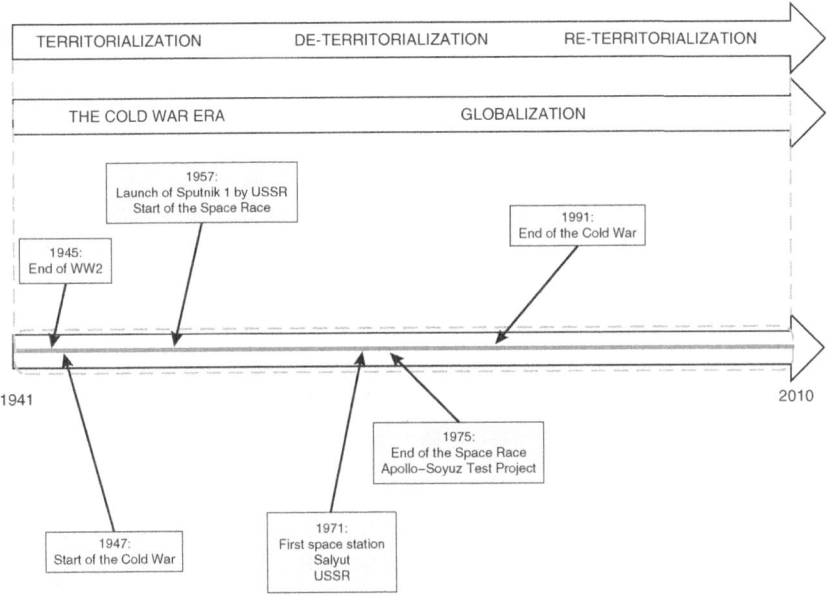

Figure 1.3 Territorialization, de-territorialization and re-territorialization

The 1990s was to be a landmark decade for collaboration in the space industry. In 1991 Cosmonaut Sergei Krikalev became the first Russian to fly aboard a US space shuttle. The following year, after traveling aboard a Russian Soyuz spacecraft, US astronaut Norman Thagard, along with Cosmonauts Vladimir Dezhurov and Gennady Strekalov, spent 115 days on Mir. This event laid the foundation for future collaboration among nations, the most obvious manifestation of such collaboration occurring later in the 1990s when the first stage of the International Space Station was launched. The millennium has witnessed the continuation and expansion of the collaboration experienced during the previous decade. On 2 November 2000, the crew of Expedition One, astronaut Bill Shepherd and cosmonauts Yuri Gidzenko and Sergei Krikalev, docked at the International Space Station (ISS). They were the first people to take up residence on the ISS, staying there for several months.

Global investment in satellite navigation systems was extensive in the 2000s. The US with the GPS system, Russia with its redevelopment of GLONASS (with the help of India), China developing BeiDou, and the European Space Agency with the Galileo Positioning System, mentioned above, have all emerged as players in this sector of the industry.

Today, the space industry is still largely dominated by states. However, the United States and Russia are no longer the only 'super-powers' in the industry; Europe, India, China, Japan, Brazil, Iran and other states have joined the space race. During the 1990s, the commercial space industry began to flourish, and ties to the military lessened. As well as 'super-powers' we also have 'supermarkets'. In particular, the space market has expanded into new niche sectors: space tourism and travel, mining of resources, manufacturing opportunities, satellite technology all represent a shift toward privatization of the sphere. The new century is an important time in the history of space, not just for science, but in the opportunities it offers for business enterprise and commercialization.

States are still the major players at the time of writing, however, and continue to cooperate, the most prominent example being the International Space Station. The United States, Russia, Canada, Japan, and the European Space Agency (ESA) all contributed to the station's construction.

All the agencies mentioned previously have been government-owned and operated, but the first decade of the twenty-first century saw another transformation of the space industry with the entry of private-sector players. There is a shift emerging, from projects that are completely government-funded, to privately-funded projects. According to Suzuki (2007), during the twentieth century investment in space technology was at the infant stage and needed to be boosted by enthusiasm. Dreams and visions helped people to support a significant amount of investment. However, twenty-first century space activity will be quite different from what it was before, and must align with new social values. The social values of the twenty-first century are not just environmental and humanitarian; they also include efficiency of investment or, in other words, 'value for money'. In today's world, the financing of space business is quite different from that of 1957. Private actors are beginning to invest in space and wealthy individuals are paying for their tickets to travel into space, while national governments face severe constraints on their spending policies.

On the one hand, the globalization of financial markets, the introduction of the single currency in Europe, and neo-liberal market-oriented policies have imposed a very narrow choice of policies on governments wanting to spend their budget. On the other hand, it is no longer necessary for a person with a dream of going into space to be a 'national' astronaut; these days, s/he needs to be a millionaire. Also, private actors are investing in satellite systems through Public–Private Partnership (PPP) schemes in Europe (Brocklebank *et al.* 2000), and in the transportation system through the Commercial Orbital Transportation System (COTS) framework in the USA (Sawamura *et al.* 1992). The role of states and national space agencies is to adapt to this new social value of the efficiency of investment. It is not the state that responds to people's dreams, but the market and private capital. Investment in space therefore needs to be more responsive to social needs because it needs to return benefits to taxpayers. Today, not all taxpayers appreciate the 'progress' and 'dream' aspects of space flight, but almost all taxpayers benefit from a better environment and safer navigation.

Thus, according to Suzuki (2007) it is imperative for everyone involved in space to recognize and understand that the name of the game has changed. Space activities need to adjust to the values of the twenty-first century, including 'value for money'. Those who are keen to go into space and believe in 'progress' can no longer depend on state-sponsored space activity. After all, many of the latest technologies and progressive ideas have been realized through market interactions. Space is becoming one of them.

Along with 'supermarkets' we also have 'super-individuals', as space travel is the final frontier for entrepreneurs. Before the twenty-first century, manned spaceflight was the preserve of government agencies and their contractors in what used to be called 'the military'. As the US aerospace journalist Michael Belfiore recounts in his book, *Rocketeers*, 'a motley crew of business adventurers are investing hundreds of millions of dollars in private spacecraft' (Foust 2003). The frontrunner among the private entrepreneurs was Burt Rutan, a designer of innovative aircraft since the 1960s. On 21 June 2004, his SpaceShipOne became the first privately-funded craft to enter space (which starts at an altitude of 100km, according to the internationally accepted definition). Scaled Composites, Rutan's company, is building the fleet that Sir Richard Branson's commercial space venture Virgin Galactic will use to carry people into space for around a 5-minute journey at a cost of US$200,000 each. It also has financial support from Microsoft (Branson 2006). Aabar investments, the Abu Dhabi state-linked investment fund, has a 32 percent holding in Virgin Galactic. Many space entrepreneurs grew up at a time when it seemed reasonable for boys to assume that manned flights to the moon and beyond would be routine by the twenty-first century. They feel cheated by the way things have turned out and now wish to use their wealth to make space tourism viable while they are still around to enjoy it. As SpaceShipOne's creator, Rutan says, private enterprise, not government funding, will conquer the final frontier. He presented his vision at cnnmoney.com in an interview given on 24 February 2006 (http://money.cnn.com/magazines/business2):

> Entrepreneurs have always driven our technical progress – and, as a result, our economy. They tend to be more innovative, more willing to take risks, and more excited about solving difficult problems. They seek breakthroughs, they have the courage to fly them, and they know how to market them. They will now provide the solutions and the hardware needed

to enable human spaceflight with an acceptable risk – at least as safe as the early airliners. (Rutan 2006)

Entrepreneurs are thus embodying the great pioneers who enable humankind to advance. Many of them are persuaded by Rutan's words. They fund prizes to stimulate research. For example, the Orteig Prize was a US$25,000 reward offered on 19 May 1919, by New York hotel owner, to the first allied aviator(s) to fly non-stop from New York City to Paris or vice-versa. On offer for five years, it attracted no participants. Orteig renewed the offer for another five years in 1924, when the state of aviation technology had advanced to the point where numerous competitors vied for the prize.

In more recent times, the Ansari X PRIZE was a space competition in which the X PRIZE Foundation offered a US$10,000,000 prize for the first non-government organization to launch a reusable manned spacecraft into space twice within two weeks. It was modeled after early-twentieth-century aviation prizes, and aimed to spur development of low-cost spaceflight. The prize was won on 4 October 2004, the 47th anniversary of the Sputnik 1 launch, by the Tier One project designed by Burt Rutan and financed by Microsoft co-founder Paul Allen, using the experimental space plane SpaceShipOne. US$10 million was awarded to the winner, but more than US$100 million was invested in new technologies in pursuit of the prize. The fourth X PRIZE was announced in September 2007. Google founders, Sergey Brin and Larry Page, are using company money to fund this, the fourth X-PRIZE, to create a private race to the moon (X PRIZE Foundation 2009). The challenge calls for privately-funded teams to compete in successfully launching, landing, and then traveling across the surface of the moon while sending back to Earth specified photo and other data. The X PRIZE will award US$20 million to the first team to land a robot on the moon that travels more than 500 meters and transmits back high-definition images and video. The X PRIZE US$20 million first-place prize is on offer until 31 December 2012; thereafter it offers US$15 million until 31 December 2014. NASA has started to award prizes under its Centennial

Challenges scheme to spur technological development to enable lunar exploration, while in the same vein the Northrop Grumman Lunar Lander Challenge has been running since 2006.

Globalization and the pivotal role of the space industry

'Rapid transportation from one continent to another and the growing capabilities enabled by information and communications technology have led to the emergence of a kind of global consciousness' (Dudley-Flores and Gangale 2007). 'Its emergence has been aided by space exploration' (Robertson 1992). Globalization has been aided by the space industry since the industry's inception. Ground-breaking technological developments that have arisen out of the industry such as satellite-based commodities have revolutionized weather prediction, television entertainment and navigational devices. Space has certainly been an influencing factor in 'bringing the global local' (Castells 1996). However, this relationship is not mutually exclusive. Without globalization it would not have been possible for the space industry to take off, so to speak. This is evident by looking at some of the fundamental driving forces behind the phenomenal process that is globalization:

- Increased expansion and technological improvements in transportation and communications networks;
- Liberalization of cross-border trade and resource movements;
- Development of services that support international business activities;
- Growing global consumer demand for products (space tourism, satellite television, GPS, etc.);
- Increased global competition and collaboration; and
- Changing political and economic situations.

Without the influence of the above forces, the space industry might not have evolved at all, let alone reached the level of success attained today. Each of these driving forces is necessary to maintain the industry, and to aid in its growth and expansion. A similar symbiotic relationship also exists between globalization and the world of international business. It has been suggested that the world is 'flat' in the sense that globalization has leveled the competitive playing field between the developed and emerging countries (Friedman 2005). The cross-boundary connections between suppliers and markets, for which globalization is renowned, would not have happened without international business. This similarity is a crucial factor in determining whether space is the new frontier of international competition. Space, like international business, survives on the fruits of globalization, and to be without it would render global competition obsolete.

The importance of the space industry for the current process of globalization has been emphasized by Dudley-Flores and Gangale (2007). By taking an 'astro-sociological approach' the authors claim that, as an instrument of the Cold War, the satellite helped to bring about the end of it by speeding up the process of globalization across several broad categories of interactive phenomena: information technology, ecological effects, social movements and organizations, concern for equal rights, global recognition, the quest for breakthrough ideas, and economic growth. These drivers were previously identified as the key patterns of interaction driving the globalization process (Peterson 1972).

Information technology

According to Lebeau (2008), space technology arrived in the 1950s from an unexpected quarter – not from the USA, which had been heralding its coming, but from the USSR, which had been secretly fashioning it within its armories. It appeared with a suddenness that created a shock wave in both public opinion and political circles. Since its birth, it has demonstrated great symbolic power, which is still capable of hindering perception of the social stakes attached to its development. There is, furthermore, a strong tendency, according to Lebeau, both within professional circles and among the general public, to think of space as endowed with its own character, one distinct and separate from all other human activities. Governments have been quick to exploit these characteristics for their own, usually short-term, ends. The 'moon race' and the use of space to glorify

Soviet ideals were both examples of this. All the technologies that, over the course of history, have widened the field accessible to mankind have been marked by this dream element. However, in the long term, what controls the development of a particular technology and determines the field in which it will be applied, is the way it takes root in society through the services it provides and the capacities it creates.

The computer has been heralded as the landmark invention of the advanced industrial way of life (Peterson 1972), but Dudley-Flores and Gangale (2007) argue that satellites and all they can do in Earth's orbit have provided much of the impetus behind advances in computer technology. They highlight that computers were necessary to guide the rockets that were the satellites' delivery systems; they were needed to track satellites; and they were needed to process the large amounts of data that came from satellites. They conclude that computer and the satellite are the heart and soul of information and communications technology (ICT).

According to Dudley-Flores and Gangale (2007), among the factors that drive the globalization process, ICT is the most seminal, because it increases the frequency of human interactions at an exponential rate. Since the speed of social change is itself partly a function of the speed and ease of these interactions, they argue that the rapid exchange and processing of information has contributed to the global erosion of existing hierarchical structures. Hierarchical structures are the hallmark of tribalism, nationalistic movements, entrenched governmental bureaucracies and most corporations (Castells 1996). In the inevitable reordering of the system following the end of the Cold War, Dudley-Flores and Gangale (2007) consider that the 'winners' were those societies that had a more open stance toward globalization, with the old fear of mutually assured destruction that characterized the Cold War giving way to the uncertainty of the re-negotiation period. They see ICT and all the other things that drive the globalization process as breeding the new social forms that will make up the re-negotiation of global civil order. According to Dudley-Flores and Gangale, this world order, in the decades and centuries to come, will find itself inevitably extending off the planet. But what does this mean for space today? Are we no longer interested in space? Have space activities lost their function as the symbol of statecraft and the glory of technological advancement? According to Suzuki (2007), to some extent this is the case. Space activities can no longer be sustainable if we use them only for national prestige and as a marker of 'progress'. Suzuki (2007) notes that some countries, such as China, South Korea and many developing nations, still cling to the notion of progress and national prestige, just as many industrialized countries did in 1957, and that these so called 'latecomers' are initiating their space programs as the 'old comers' did during the 1950s and 1960s. In contrast, he argues that, for the 'old comers', space activities need more justification than progress and national prestige, and need to serve society and its values. Suzuki maintains that the social value of space lies in its ability to solve the dire problems of society and to help provide the infrastructure for the solutions. Thus, environmental monitoring, disaster management, support for navigation, long-distance communication and enhancing security have become prioritized objectives for space activities. Thus, for Suzuki, 'Progress' is not just the progress of technology but also that of humanity. In other words, according to Suzuki, space is no longer an end in itself, but a tool to achieve social objectives.

This is the rationale for the space activities of the twenty-first century. During the twentieth century, there was investment in space technology because it was at an infant stage and needed support to become established. However, as noted earlier in the chapter, twenty-first-century space activity is very different from what it used to be, and must now align itself with the new emerging social values.

Those who are keen to embrace the space adventure and believe in 'progress' cannot depend on state-sponsored space activity. After all, many of the most innovative technologies and ideas have been realized through market interactions. Space is increasingly becoming one of them.

Ecological effects

Although we are still in modern times, Suzuki (2007) maintains that while human beings are no longer in pursuit of progress and dreams of a high-tech, science-fiction life, they still use space technology to solve problems on Earth and to improve their quality of life despite limited financial resources. According to Suzuki (2007), the modernity in which we are living now is different from the modernity of 1957. None the less, he argues that this new modernity (or 'high modernity') was only made possible by those who contributed to the translation of the 'dream' technology into reality, and that, without these achievements, people would not be able to enjoy their present quality of life and appreciate the importance of the environment.

Dudley-Flores and Gangale (2007) argue that imaging the planet, a direct product of space exploration, has enabled a larger awareness of the biosphere and that conservationism flowered into modern environmentalism as a result of imagery from space, and other instruments and processes of space research and development.

Dudley-Flores and Gangale (2007, p. 3) claim that 'Space exploration does not stand apart from the Globalization process. It is part and parcel of the thing it has magnified.' The globalization process therefore requires us to re-think the exploration of space. The fast-growing emerging economies of China and India are ramping up their space activities at a very fast rate. It remains to be seen whether they will collaborate with others as the former adversaries of the Cold War, the USA and the former Soviet Union, are increasingly doing today, or will compete in the manner of the Cold War adversaries. Onoda (2008) describes how the development of remote-sensing technology, for example, was greatly spurred by wartime needs and the international settings of the Cold War, and thus historically it is associated with concerns over national security and sovereignty. Nevertheless, the technology also allowed humans to view the entire Earth, and has enabled global data-gathering for environmental policy-making purposes (Johnson *et al.* 2008).

Similarly, Dudley-Flores and Gangale (2007) note that there is an historic trail to collaboration among countries, extending to increasingly longer-duration space missions. They argue that mastering long-duration space exploration is a prerequisite to a permanent human presence in space, which is nothing short of the expansion of human ecology away from the Earth. But they note that, on the verge of longer-duration missions, as in a manned mission to Mars by the USA, conceptualized for the 1980s, events of American history intervened – namely, at the end of the 1960s in the form of the decisions made by the Nixon Administration to cut space funding. In the meantime, considerations related to the planet and its survival, such as securing energy sources and climate change, have assumed saliency. The emergence of such critical issues poses the question of how 'globalized space' can respond to them.

According to Dudley-Flores and Gangale (2007), the significance of the inquiry into the globalization of space relates to whether space exploration will be able to play a large role in addressing these issues, leading to the expansion of human ecology on the Earth and in places away from the Earth. By making progress around issues related to energy, climate change and other mammoth challenges facing the planet, they believe that space can become demonstrably relevant to a wider global audience interested in survival. As Stephen Hawking has recently asserted, space offers the key to human survival (Hawking 2002). He contends that, since war, resource depletion and overpopulation threaten the existence of the species as never before, the only chance of long term survival for the human species is not to remain inward-looking on planet Earth but to spread out into space 'If we want to continue beyond the next hundred years, our future is in space' (p. 63).

Social movements and organizations

The information and communications technologies developed by the space industry are essential tools to enable social movements and organisations to meet their goals today. Dudley-Flores and Gangale (2007) speak of the 'organic' growth of non-state-actor organizations and note that most commentators on globalization have remarked (see, for example, Castells 1996; Vallaster 2005), that much about the social formations of our modern world is characterized by 'network' structures diffused from both the biological world and the worlds of broadcasting and the World Wide Web.

It is possible to argue that the power-shift from nation-states toward regional/global political or economic institutions, and the lack of or weak democratic control over these 'higher' levels of governance, has prompted civil society organizations – and more specifically social movement organizations – to organize themselves beyond the nation-states in order to critically question the legitimacy and policies of international economic and political actors (Cammaerts 2005). According to Cammaerts, transnational civil organizations allow citizens to link up with a community of interest and action beyond their own nation-state. As such, transnational civil society could be perceived as resulting from 'globalization from below', an attempt to counter-balance the globalizing economic, political and cultural spheres that increasingly escape the sovereignty of the nation-states. It is within this complex political context that the use of ICT by transnational social movements should be situated. Cammaerts' (2005) study of the different usages of interactive communication technologies by social movements has identified three main categories of use. First, social movements use ICT to organize themselves and to interact with their members, sympathizers and core staff. Second, use relates to mobilization when ICT is used to lobby within formal politics or to foster social change through online as well as offline direct actions. Third, there is the potential to strengthen the public sphere via the mediation of political debate. Here, the internet is considered by many scholars to be a potential means to extend the working of transnational social movements geographically, to organize internationally, to build global or regional coalitions with like-minded organizations, to mobilize beyond their own constituencies, and to spread information independently on a global scale, thereby supporting the development of global or transnational public spheres.

Concern for equal rights

The number of issues requiring global solutions has also increased and become more prominent on the political agendas of citizens, civil society organizations and (some) governments. Examples of such issues are: child labor, ecology, security, mobility, migration and human rights. Dudley-Flores and Gangale (2007) argue that, since communication today permeates national boundaries, there is awareness among people across the world of the living conditions of others. This awareness of the realities of others has been intensified via satellite broadcasting and the internet, which have brought images of a hard reality, making what Marshall McLuhan called in the 1960s 'the global village' (McLuhan and Powers 1992) a reality. According to Dudley-Flores and Gangale (2007), the global village has never been so real as it is now, after the first decade of the twenty-first century.

Global recognition

While global recognition was once reserved for nation-states and rarely others, Dudley-Flores and Gangale (2007) argue that it is now being extended to the individual. Since modern communication and transportation have made available the teachings and technologies of the world's cultures to nearly everyone, they consider that customers will not only stop and compare these, but, most important, they will also compare notes with others. And so individuals as well as corporations and nation-states will be empowered as a result (Friedman 2005).

According to Cammaerts (2005), this change reflects the shifting notion of citizenship. The citizenship notion has also evolved considerable since the Greek city-states or the formation and consolidation of the Westphalian nation-states. While citizenship is theoretically, but also empirically, still very much linked to and conceptualized within the 'boundaries' of the modernist nation-state; the increased globalization of the world economy; revolutionary innovations in communication, transport and mobility; ecological and demographic pressures; and ethnic and nationalistic forces have considerably undermined the sovereignty and legitimacy of that nation-state, the core of the bounded notions of citizenship. These social, economic and political transformations would suggest that it is fair to conceive of citizenship as being more complex and diverse than a classical understanding linked to rights and nationality. In political theory, this is exemplified by the emergence of several concepts of citizenship that could be called unbounded and go beyond the nation-state. Examples of these are ecological citizenship, net.citizenship, transnational citizenship, cosmopolitan citizenship, or denationalized citizenship.

The quest for breakthrough ideas

According to Dudley-Flores and Gangale (2007), the infrastructure that has spread from the satellite and the computer is 'the Gutenberg Press of our time' –and it was all made possible by the human exploration of space.

In a recent article on the future of human civilization (Shapiro 2009), Shapiro claimed that a process is well under way in which the scientific, technical and cultural information vital to human society is stored in digital form within a limited number of computer facilities. This practice is vulnerable to a variety of catastrophes, which would destroy humankind's knowledge base in addition to the losses caused to populations and structures. Shapiro (2009) believes, however, that the construction of a substantial lunar base as part of a program to ensure the survival of human civilization on Earth is a goal that would link and justify two main purposes. First, he sees the need to preserve our cultural heritage. Recent decades have seen such an explosion in the production of scientific data and cultural material that of necessity they are being stored in digital form. Older materials are also being converted to digital form, allowing much of humanity access to a treasure trove of science and art that can readily be explored and utilized in innovative ways. However, this new storage medium is more fragile than paper, both because of its inherent nature and because of its greater vulnerability to local disasters and global catastrophes. If the human cultural heritage were to be substantially damaged or lost, civilization as we know it at present could not function, and humanity would be reduced to a barbaric state. A measure of protection could be provided by using the moon for the purpose of preserving scientific and cultural documents and objects that support human civilization. Second, he sees the construction of a substantial lunar base as part of a program to ensure the survival of human civilization on Earth also providing a transcendent purpose. The Apollo program that delivered human beings to the moon a generation ago may have been a by-product of competition between nations during the Cold War era, producing media coverage and images that were inspirational. However, Shapiro emphasizes that no further purpose emerged from that presence to stimulate the imagination of the public, and no further human expeditions beyond Earth's orbit have been launched since that time. In January 2004, President George W. Bush announced his Vision for Space Exploration, which involved a return of human beings to the moon. More recently, however, President Barack Obama has revamped this vision to exclude human beings landing on the moon. Economically emerging nations have also indicated an interest in lunar exploration, but the reasons provided have not truly justified the expense involved. In the absence of a transcendent purpose, the prospects for human expansion into space remain uncertain. Shapiro believes that a unique opportunity has arisen to link two worthy causes that have emerged (that is, the need to preserve our cultural heritage and to plan a strategy for the survival of the human race) in the recent past; each of which might flounder if allowed to proceed separately.

According to Dudley-Flores and Gangale (2007), another breakthrough idea coming from the space industry is nanotechnology and this represents the logical extension of the miniaturization effort that began in the early days of space exploration.

Extra-globalization and the creation of a solar system economy

In recent decades globalization has become associated with economic growth. Dudley-Flores and Gangale (2007) argue that only an interdependent global economy could provide the capital mass or sufficient finance to support the application of breakthrough ideas, truly effective global organizations, the enhancement of each individual, the assessment of environmental degradation and climatic shift of the whole planet and the taking of action to reverse these where possible, and to extend human ecology to other locations.

They consider that when pockets of people begin to live sustainably away from the Earth, an *extra-globalization* process will begin to occur. If globalization implies the increasing interconnectedness of Earth's societies and alterations in the space endeavor, they consider that extra-globalization is the extension of these intertwined phenomena to those sustaining themselves indefinitely away from the Earth, and the dialectic set-up between them and the Earth.

Dudley-Flores and Gangale (2007) look forward to a distant time when low transportation costs make the extraction of primary resources from space economically feasible. They noticed that, in the past, colonization led to exploitative relationships, in which colonial powers took advantage of cheap, unskilled labor in the colonies to extract raw materials and these trade relationships have persisted into the twenty-first century. Wallerstein (1980) refers to this trade system by distinguishing its main components: the industrial core of the capitalist world-system, the semi-periphery of lesser industrialized states, and the underdeveloped periphery. However, as Dudley-Flores and Gangale (2009, p. 1) notice, these 'space settlements will be high-tech by their very nature, and will be populated by highly-educated, highly skilled workforces. Thus, once settlements are able to provide for their own subsistence, they will be able to turn to high value-added productive activities, many of which will be competitive with terrestrial products elsewhere in the solar system due to gravitational advantage'.

According to Dudley-Flores and Gangale (2009), should these settlements develop on the moon and on Mars, and even elsewhere, the most credible use of space resources would be either for local use or use elsewhere in space. In the absence of commercial revenue from direct trade with Earth, interplanetary trade not involving Earth directly would necessarily rely on government contracts to private companies to provide goods and services to government-owned operations. In this context, while Earth is the source of investment capital for outer space, the accumulation of hard currency in outer space will require the creation and exchange of something that is of value to

Earth. These productive activities will form the basis of a largely commercial, non-governmental, solar-system economy.

As commercial interplanetary trade develops, Dudley-Flores and Gangale (2009) consider that interesting avenues of speculation might arise. For example, they draw on Adam Smith ([1776] 1986) and the theory of absolute advantage to propose that, if all the planets could specialize in producing the commodity in which they all enjoy absolute advantage, then all would profit from trade. Each planet could specialize in the production of only that commodity for which it enjoys absolute advantage while buying the cheaper commodities from the other planets. Thus, as a result of this specialization, pr\oductivity would increase on all the planets. Table 1.2 summarizes Friedman's arguments characterizing globalization, and the main changes that have taken place throughout the evolution of the business of space.

Table 1.2 Characterizing globalization and the business of space

Friedman's arguments characterizing globalization	The business of space
Collaboration	1975: The Apollo–Soyuz Test Project saw collaboration between the US and the USSR 1986: The Mir space station hosted astronauts of several nationalities 1998: The International Space Station (ISS) saw the collaboration of the USA, Russia, Europe, Canada and Japan
Many super-powers	Originally: USA versus USSR 2010: Europe, India, China, Japan, Iran, Brazil
Supermarkets	Space tourism and travel Mining of resources Manufacturing opportunities Satellite technology
Super-empowered individuals	2004: Rutan Ansari X PRIZES

2

Industry analysis

Introduction

We have so far described the evolution of space-related activities from the Cold War era to the globalization era. Much of the twentieth-century space activity was dominated by Cold War considerations, reflecting super-power rivalries and representing one more source of competition between the USA and the Soviet Union. The era of globalization has been reflected in the evolution of space-related activity. Competition in space has been accompanied increasingly by cooperation reflecting the collaborative imperative of globalization. Just as previously state-controlled activities in many spheres have been privatized, so have space-related activities entered the private realm. Increasingly, space-related activities have become the business of space as entrepreneurs and other private entities not only embark on pursuits previously only undertaken by governments and their agencies, but also begin to venture into previously unexplored and unexploited opportunities in space. These developments are consistent with the dialectic of de-territorialization and re-territorialization. More obviously, the establishment of permanently manned space stations and the vision of establishing lunar and Martian colonies are also consistent with the dialectic of de-territorialization and re-territorialization. As space activities continue their evolution from space race to the business of space, the globalization perspective provides us with a useful means of framing this evolution, and with an enhanced understanding of the dynamics of the business of space.

Industry definition

Once we had defined the evolution of the sector within the context of globalization, we were then presented with the task of defining the space industry. It became apparent that many articles and experts who analyze the 'space industry' do not in fact define the industry at all. The term is used so widely, and so loosely, that one wonders whether the term has any real meaning: 'The usual problem with the phrase 'space industry' is that it is too inclusive: it encompasses any number of companies for whom space may not necessarily be at the core of their business, in an effort to make the industry look as large, and thus as prominent, as possible' (Foust 2003).

During our research, an investigation into what the space industry entails uncovered the reasoning behind this inclusiveness. Many of the support activities that would be involved in an aspect of the 'space industry', such as wireless communication, allow for industry analysts to show greater benefits and rewards from the industry as a whole. As is well known, every industry has a collection of organizations, each serving the needs of particular groups or market niches, so wireless communications can therefore be placed in an outright industry of its own.

In broad terms, the Organisation for Economic Co-operation and Development's (OECD's) Global Forum on Space Economics defines the space economy as: 'All public and private actors involved in developing and providing space-enabled products and services'. It comprises a long value-added chain, starting with research and development (R&D) actors and manufacturers of space hardware (for example, launch vehicles, satellites, ground stations) and ending with the providers of space-enabled products (for example, navigation equipment, satellite phones) and services to final users (for example, satellite-based meteorological services or direct-to-home video services) (OECD 2004).

Thus the space economy is larger than the traditional space sector (for example, rockets and launchers); and it involves increasing numbers of new services and product providers (for example, geographic information systems developers, navigation equipment sellers) who are using space systems' capacities to create new products. Hence, it must be noted that when we refer to the space industry, we are in fact referring to all economic activities in relation to space.

Estimates of the size of the space economy vary considerably, because of a lack of internationally comparable data. Worldwide, institutional budgets (around US$47 billion in 2005 for OECD countries) and new commercial revenues from space-derived products and services (around US$110–120 billion in 2006) indicate that the underlying trend in the space economy is one of growth (OECD 2007). And this remains true, despite the cyclical nature of commercial space activities (for example, regular replacement of telecommunication satellite fleets).

Figure 2.1 provides a simplified view of the space economy; a public or private actor may be involved simultaneously in several space activities (for example, being a manufacturer as well as an operator and service provider).

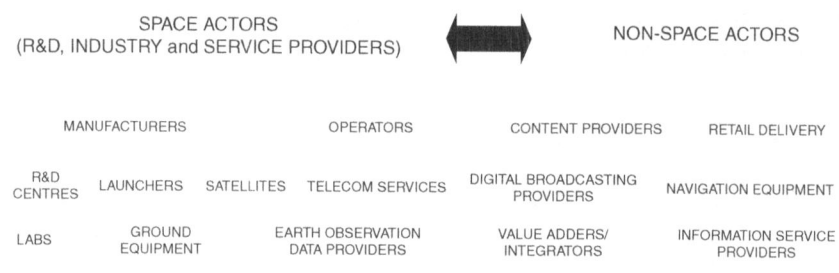

Source: OECD (2007).

Figure 2.1 Overview of the space economy

Governments play a key role in the space economy, as investors, owners, operators, regulators and customers for much of its infrastructure. As in the case of other large infrastructure systems (for example, water, energy), government involvement is indispensable to sustain the overall space economy and to deal with the strategic implications of such complex systems. In the case of space, the infrastructure can be used for both civilian and military applications, as space technologies are by their nature dual-use, and military developments often pave the way for the development of civil and commercial applications (for example, today's rockets are derived from missiles).

In particular, we refer to the space industry as the relationship between the space mission and the **R&D** sector, which, through its scientific explorations, gives rise to limitless economic potential. Figure 2.2 illustrates this process.

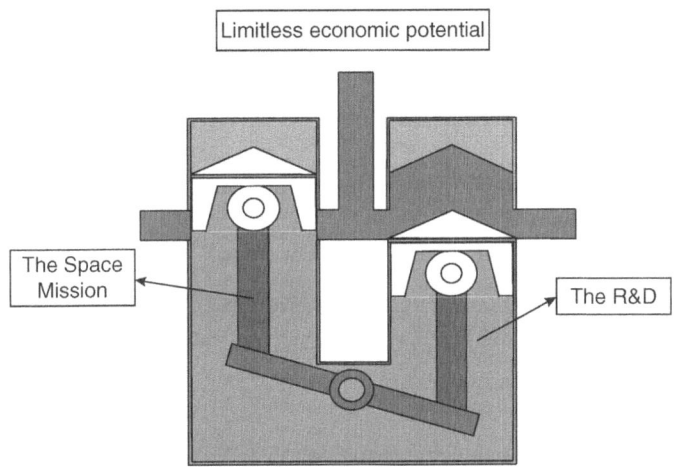

Source: Taken from http://s210.photobucket.com/albums/bb313/acurafan07/?action=view&current=drawing-1.jpg.

Figure 2.2   The space industry and its core components

The first component of the definition, the space mission, refers to the phenomenon of the wondrous endeavor to conquer the unknown that has always been with humankind, but which today is embodied by the move to be in the outer reaches of space. That mission is sponsored by national governments and private corporations, and serves a number of areas such as the military and telecommunications. The second component – the **R&D** sector – is the intellectual transformer of all of the efforts of the space mission. It provides the knowledge drawn from the collected data and produces new techniques, understandings and technological devices. Figure 2.3 provides a breakdown by country of **R&D** expenditure in the space industry.

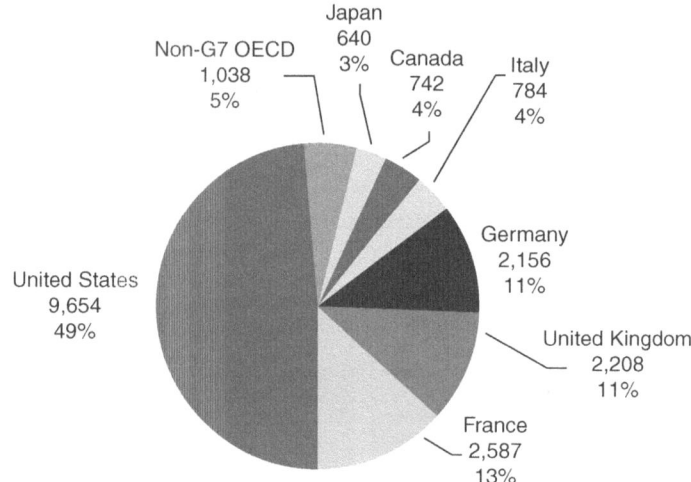

Source: OECD (2007).

Figure 2.3 R&D expenditure in the space industry for OECD countries, 2002 (in millions of US$ using PPP and percentages of OECD aerospace R&D total)

The relationship between the space mission and R&D is one of mutual complementarity. The industry is a dynamic one which functions like the pistons in an engine, where one feature, the space mission, propels the R&D sector and that in turn informs, enhances and accelerates the space mission to higher levels of achievement.

Current state of the space sector

At the time of writing, the space industry is still largely dominated by states. Space budgets refer to the amounts that governments have indicated they will provide to public-sector agencies or organizations to achieve space-related goals (for example, space exploration, better communications, security). Thus for OECD countries, space budgets may serve both civilian and military objectives. However, significant portions of military-related space budgets may not be revealed in the published figures. Data for non-OECD countries – Brazil,

Russia and India – refer to civilian and/or dual-use programs. Chinese figures are only estimates and not official data. Other estimates of China's space budget (from diverse Western and Asian sources) range from US$1.2 to more than US$2 billion (OECD 2004).

Over 30 countries have dedicated space programs, and more than 50 have placed satellites in orbit, mainly for communications purposes (see Figure 2.4). According to the latest OECD report (2007), in 2005 civilian and military budgets for space programs of OECD countries totaled about US$45 billion (though data were not available for some smaller OECD countries). Of this amount, over 81 percent was accounted for by the USA, followed by France, Japan, Germany and Italy. US space budgets picked up, especially after 2001, with the 2006 estimated budget being more than 30 percent higher than five years earlier. The general trend shows US military space budgets (that is, the Department of Defense) rising as a percentage of the total, especially since 2001. Europe also budgeted significant amounts for space programs: about US$6 billion in 2005. An examination of European budgets shows that the three largest contributors (France, Germany and Italy) accounted for 76 percent of the overall European total, including 90 percent of national and 68 percent of the European Space Agency (ESA) budget totals. Several non-OECD countries have also boosted their civilian space investments significantly over the past few years and are continuing to do so. In 2005, the Russian space budget was estimated at US$647 million (18.3 billion rubles) and India's budget at US$714 million (31.48 billion rupees), with Brazil's at US$92 million (223 million reals). The Chinese budget was estimated at a tenth of the NASA budget by Chinese officials, or around US$1.5 billion, in 2005. The national public space budget as a percentage of GDP for 2005 was the greatest in the USA, at 0.295 percent, about three times higher than France. The top ten OECD countries included all the G7, except the UK. Also, the report notes that the three major non-OECD space countries – India, Russia and China – all ranked within the top five, ranging around the G7 average of 0.084 percent.

We can also divide the space sector into a supply and a demand side. The supply side can be determined as 'embracing all

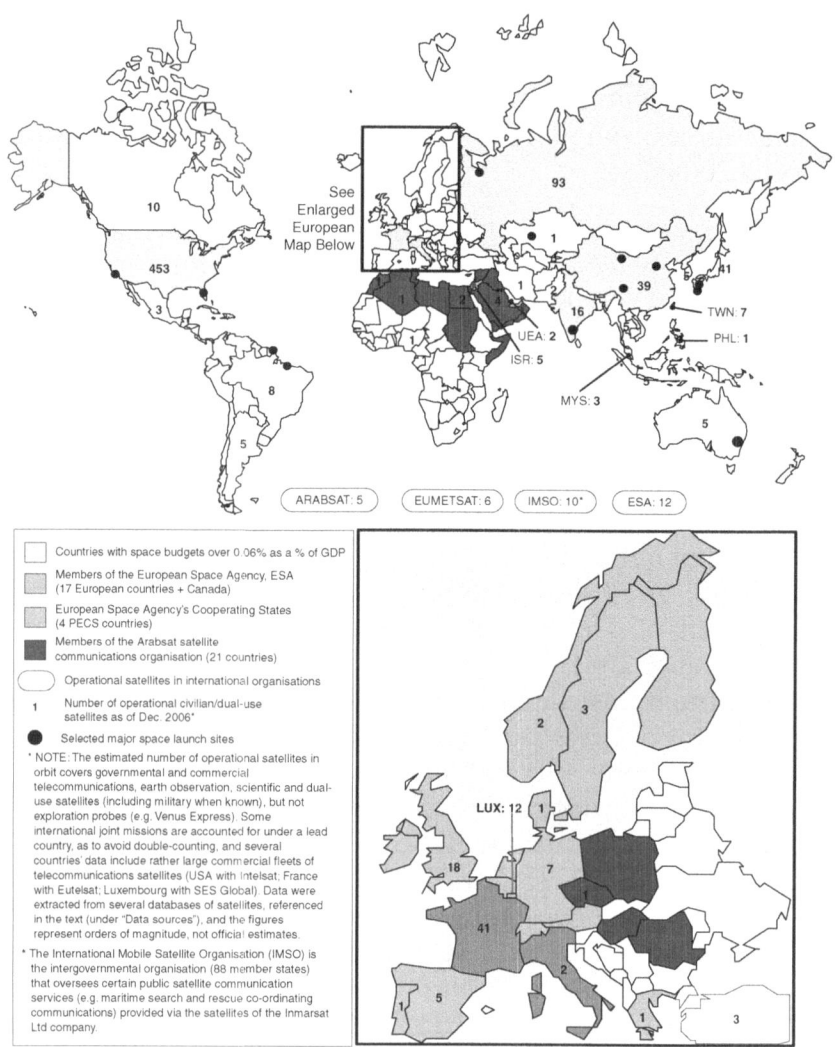

Source: OECD (2007).

Figure 2.4 Countries with operational satellites in orbit, as of December 2006 (estimates)

public and private actors involved in providing space-enabled products and services' (OECD 2007). These actors belong to a value-added chain consisting of an upstream and downstream segments. The upstream segment comprises the manufacturers of space hardware and the suppliers of launch services, while the downstream segment includes the operators of satellites and suppliers of space-enabled products and services. Space agencies play a decisive role in both segments because they conduct both upstream and downstream R&D and from time to time act as operators of space systems (OECD 2007).

The demand side includes two major elements: the institutional market and the commercial market. The institutional market obtains space assets for causes that range from manned space flight and scientific investigation to fundamental public services and supporting R&D (OECD 2007).

The commercial market refers to private or semi-private firms supplying space-based services or space-enabled products to final customers or other firms. There are three major parts of the commercial market: telecommunications (mobile and fixed services), Earth observation (EO) and location-based services (LBS). The progress of the commercial market is dependent on the development of the institutional market – that is, the commercial launcher/satellite market would almost certainly not exist in the absence of an institutional demand (Nelson and Winter 1982); OECD 2007).

The turnover in the commercial market in Europe, especially with commercial satellites, was increasing until there was a breakdown in the year 2002 following the telecom and dotcom crashes (Eurospace/ASD 2008).

Current state of the upstream segment

Launch industry

At the time of writing, a dozen countries have an autonomous capability to launch satellites into orbit. These include the USA, the Russian Federation, France, Japan, China, the UK, India, Israel, Russia, Ukraine and Iran. Many more countries have the capability to design and build satellites, but are unable to launch them, instead relying on foreign launch services. This list does not consider these countries, but only lists those capable of launching satellites indigenously.

The international space launch industry plays a pivotal role in enabling commercial and non-commercial actors to engage

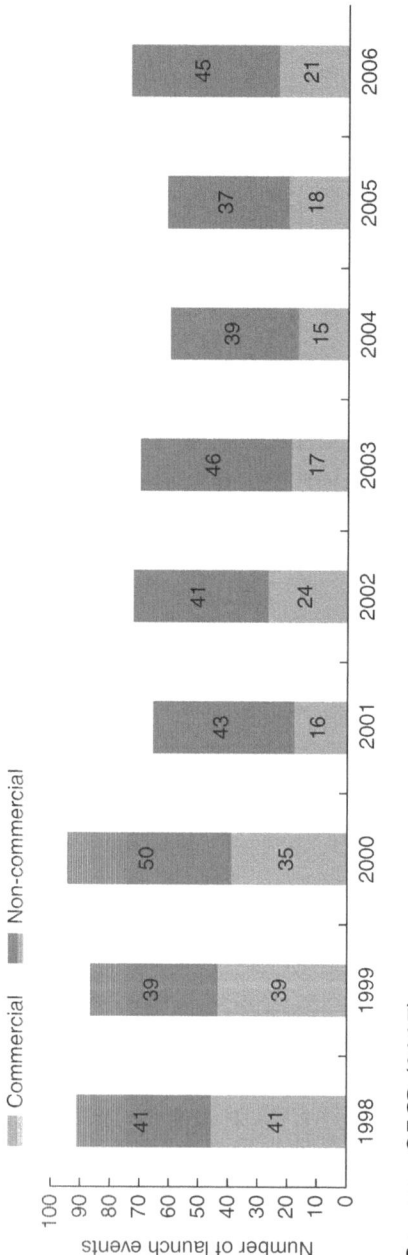

Source: OECD (2007).

Figure 2.5 Total commercial and non-commercial launch events, 1998–2006

in civilian and military space activities. Commercial launches have decreased, largely because of the financial crisis faced by telecom operators in 2001 (see Figure 2.5).

Comparing the year 2000 with 2001, there was a 30 percent decline in launches. With only 59 launches attempted all over the world in 2001, this was the smallest number since the early 1960s. The early 2000s saw the emergence of new launching capacity as new heavy launchers entered the market to satisfy the trend for larger communication satellites and the expectations raised by the dot.com bubble.

An examination of all launches by country from 2000 to 2006 reveals that, while all the major launch providers (the USA, Russia and Europe) had fewer launches in 2006 than in 2000, China had more than maintained its numbers of launches in the same period (see Figure 2.6).

An examination of commercial launch events over the periods 1996–2000 and 2001–6 reveals that Russia had a higher proportion of commercial launch events in the latter period (see Figures 2.7 and 2.8).

Revenues from commercial launches have tended to decline with declining launch activity (see Figure 2.7). The cyclical nature of satellite activities (that is, the need to renew satellite fleets) and the growing number of countries with space programs should contribute to more space launches in the decade 2011–20. International competition in commercial markets is also likely to increase.

Satellite manufacturing

Worldwide, satellite industry revenues remained relatively stable from 2002 to 2005 at around US$35–36 billion, with an increase in 2006that reached higher levels than in 2000 (see Figure 2.9).

A continuing recovery over time is anticipated, based on the cyclical nature of space activities (for example, the renewal of satellite fleets), though the growing number of actors is forcing increased international competitiveness. A breakdown of the total manufacturing revenues for this sector shows that, while

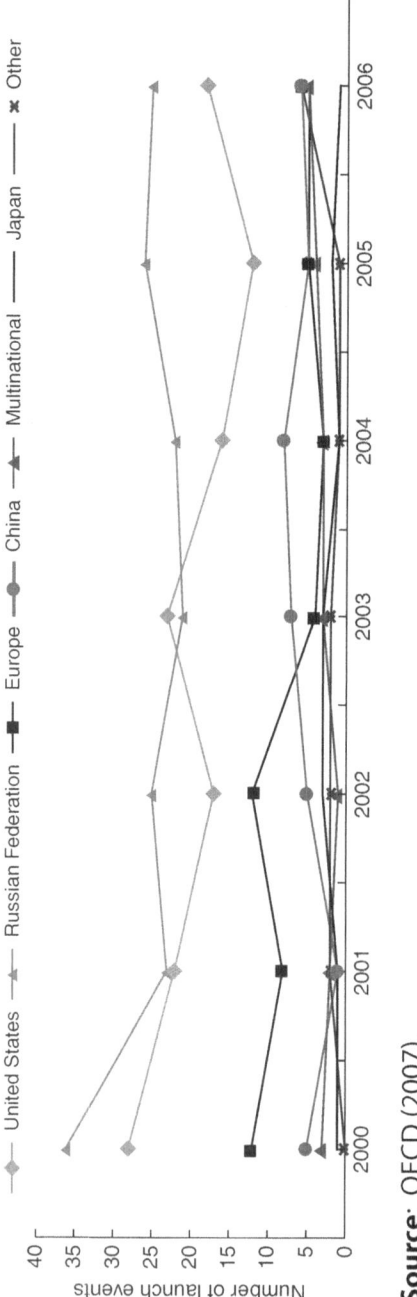

Source: OECD (2007).

Figure 2.6 Total launch events by country, 2000–6

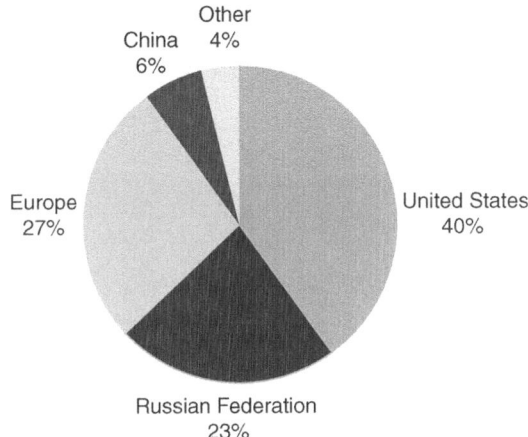

Source: OECD (2007).

Figure 2.7 Commercial launch events, by country, 1996–2000

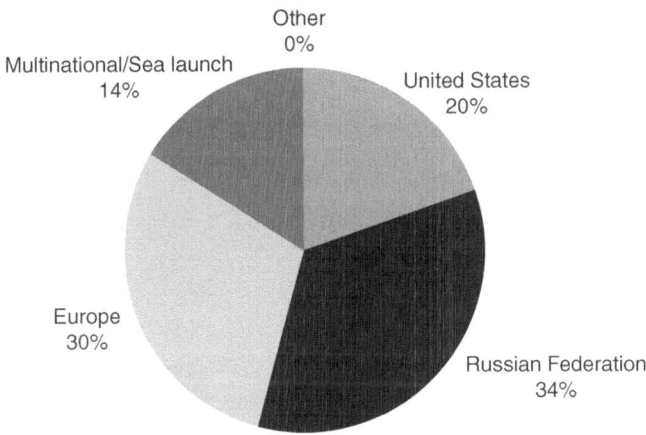

Source: OECD (2007).

Figure 2.8 Commercial launch events, by country, 2001–6

revenues associated with ground equipment grew over the period 2000–6, revenues associated with launch and satellite manufacturing areas mainly shrank (see Figure 2.10).

This trend is reflected in the rising percentage of the total revenues coming from the ground segment and a proportional

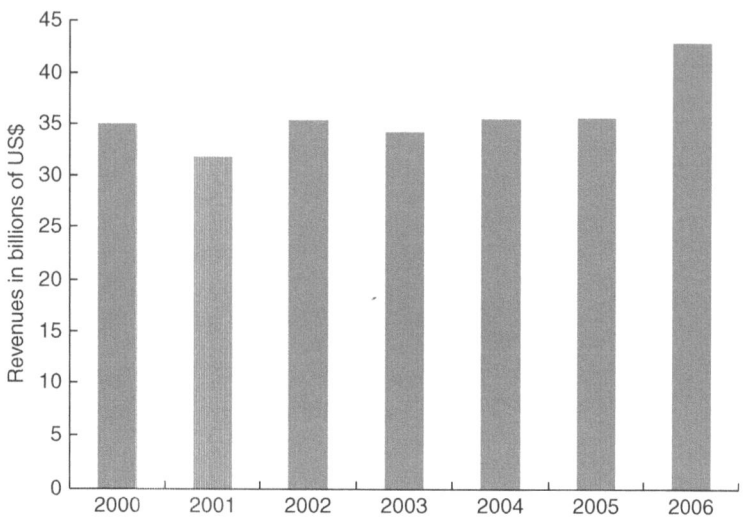

Source: OECD (2007).

Figure 2.9 World satellite manufacturing revenues, 2000–6, in billions of US$

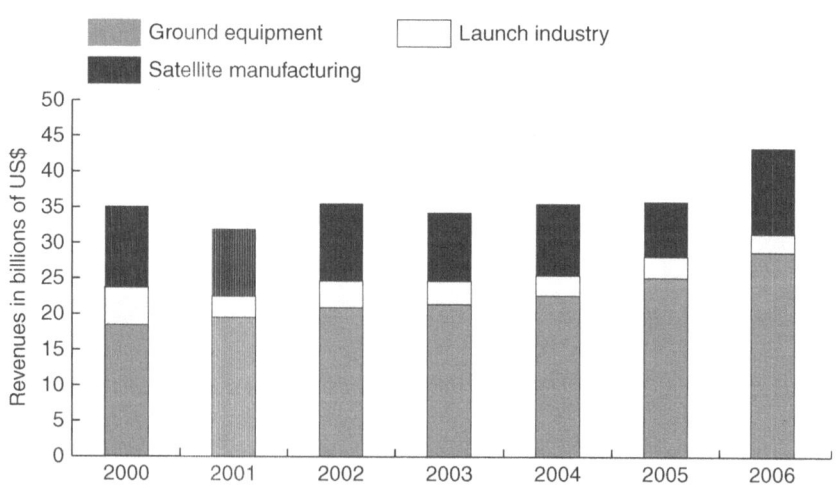

Source: OECD (2007).

Figure 2.10 World satellite manufacturing revenues, by sector, 2000–6, in billions of US$

decline in both the launch and satellite manufacturing markets (see Figure 2.11).

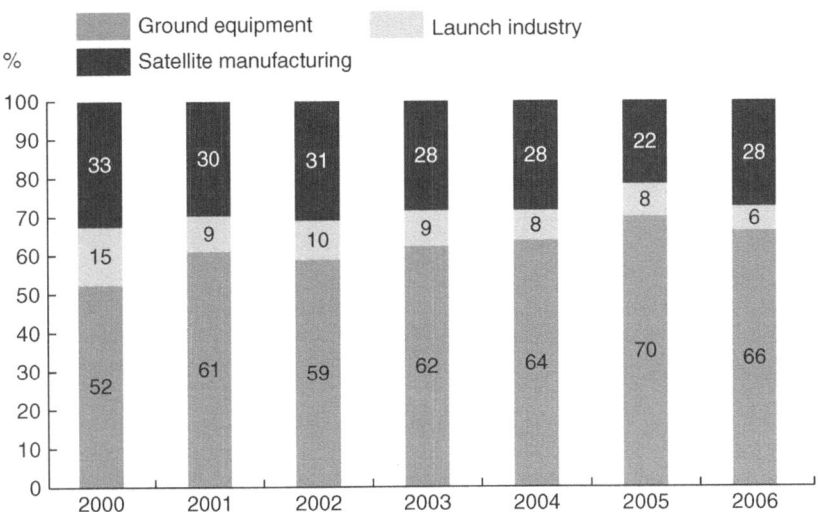

Source: OECD (2007).

Figure 2.11 World satellite manufacturing revenues, by sector, 2000–6, percentages

An examination of European space-related manufacturing units shows a similar picture, with sales decreasing since 2000, though showing some recovery in 2006. Worldwide launch revenues in 2006 had returned to the high levels of 2000, when the USA and other players were vying for launch activity (see Figure 2.12).

Satellite producers compete on price and on the quality of the offerings and features provided. Thus the barriers to entering this market are presumably lower than for the launchers market, and some companies can survive by concentrating on niche markets. As well as the launcher manufacturers, satellite producers have encountered hard times in recent years, leading to consolidation among satellite operators. In particular, the significant advancement in terms of the stability and capacity of spacecraft has led to a sharp decline in the demand for additional satellites. In 1998, for example, more than 150 satellites were launched but demand,

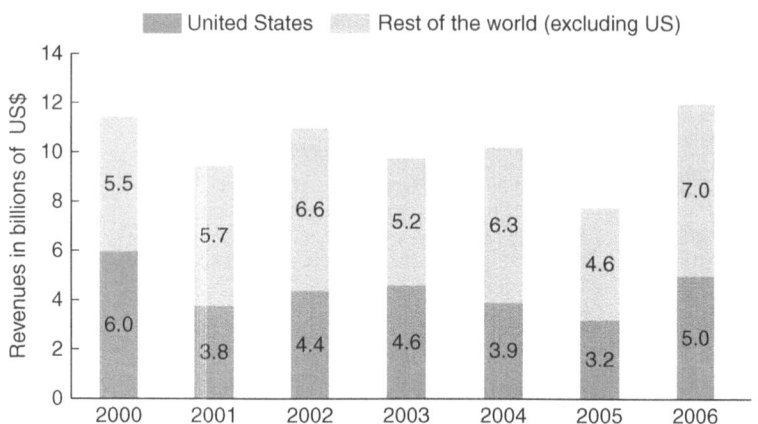

Source: OECD (2007).

Figure 2.12 Worldwide launch revenues, 2000–6, in billions of US$

especially for commercial communication satellites, fell sharply and in 2001 only 75 were launched. This is the lowest number of launches over the past decade and a 32 percent decrease compared with the year 2000. In 2002, 80 satellites were launched (OECD 2007).

In the short term, a significant turnabout in terms of new satellite orders is not expected, though some new applications are in the pipeline. For example, the internet is driving the development of business and commercial applications for commercial satellites, but the majority of firms that plan to offer satellite-based internet services will lease capacity on assets already in orbit and will not acquire new satellites, at least in the medium term (OECD 2004).

The ray of hope is the military market, as military contracts offer well-paid and long-term activity for contractors whose commercial business is contracting. But this is more of a benefit to US suppliers of military space equipment than for non-US companies. Some exceptions in Europe are projects such as Galileo and Global Monitoring for Environment and Security (GMES), which could preserve the prospective demand for satellites. There are some signs of recovery in this sector; for example, Lockheed Martin Commercial Space Systems and Alcatel Space became profitable again in 2003 after facing years of difficulties. This recovery for satellite producers was mainly achieved by cutting down costs through subcontracting. Lockheed Martin, for example, was able to downsize by reducing its workforce by 40 percent over two years. This generated new commercial opportunities for subcontractors who deal with various prime contractors. Nevertheless, the revitalization is assumed to be subdued (OECD 2007).

Resources

The mining and utilization of the abundant resources in existence on the moon and near-Earth asteroids (NEAs) could potentially harvest trillion-dollar industries in the future. The exploitation of the available resources presents multiple opportunities for financial reward. However, as one might expect, there are great barriers to overcome before the potential of space in terms of resource commercialization can be realized. These barriers are legal, scientific, mechanical and financial in nature, and all impede the progress and evolution of the space resource industries.

Space resources that attract commercial attention can be classified basically into the following:

- water;
- solar energy;
- metals – ferrous, precious and strategic; and
- fusion reactor fuels.

The exploitation of each resource presents unique combinations of opportunities and threats, and these are at various evolutionary stages in their development. Their utilization also varies; some resources are to be used in space, while others are materials commercially destined to return to Earth or to be used as a source of energy supply for the Earth.

Water

This is an extremely versatile resource with the ability to provide propellants for rocket engines as well as being an essential life-sustaining material. Experts say that the hydrogen found in ice could be used to make fuel for space exploration, and oxygen in water could provide air for explorers and even for colonies. To haul the weight of water up through the Earth's gravity well is extremely costly, and if water can be accessed in space it has the potential to revolutionize both the mechanical and human aspects of space exploration and commercialization. Subsequently this resource might assume a lucrative position of economic importance in the era of space commercialization, and billions of dollars have been spent to date in search of this precious commodity in space. The European Space Agency's Infrared Space Observatory (ISO) detected water vapor and confirmed visual observations of water in space between 1995 and 1998. NASA's Lunar Prospector spacecraft orbited the moon in the years 1998 and 1999, and Alan Binder, Lunar Prospector's principle investigator, reported that the investigation had discovered small expanses of the lunar surface that are believed to hold water-ice crystals mixed with surface materials. The Lunar Prospector's intentional crash-landing on the moon uncovered evidence of hydrogen and possibly surface water:

> there could be vast quantities of it – as much as 200 million tons of ice crystals buried 18 inches (45 centimeters) or so beneath the Moon's dusty surface. Thawed, that would be vast enough to fill a lake about 2 to 3 miles (3 to 5 kilometers) wide by 32 feet (10 meters) deep. (SPACE.com 2009)

Chemically broken down into hydrogen and oxygen, water could provide fuel for rockets or electrical generators, and along with breathable air, this classifies its application as in-situ resource utilization (ISRU). Gerald Sanders, Chief of the Propulsion and Fluid Systems Branch at NASA's Johnson Space Center describes ISRU as the stepping-stone for space development. If human beings ever go into space en masse, for extended periods of time, for either commercial or tourism purposes, space resources could become a necessity to sustain life. With the evolution of the space tourism industry there is in turn greater strategic and economic interest in establishing ISRU, by NASA and other programs. The amount of interest being paid to this space resource is therefore intensifying with the development of space tourism.

Solar energy

Energy consumption has been constantly increasing for many decades and has brought with it severe environmental consequences which have captured the world media's attention as solutions are sought to deal with this global problem. The OECD is a unique forum where the governments of thirty democracies work together to address the economic, social and environmental challenges of globalization. The International Energy Agency (IEA) is an autonomous body, established in November 1974 within the framework of the OECD, with the aim of implementing an international energy program. The IEA produces an annual report, the *World Energy Outlook* (WEO), and the latest report details how the trends in energy demand, imports, coal use and greenhouse gas emissions predicted up to 2030 have continued to worsen year by year. WEO identifies the emerging giants of the world economy, namely China and India, as transformers of the global energy system:

> The consequences for China, India, the OECD and the rest of the world of unfettered growth in global energy demand are alarming. If governments around the world today stick with current policies the world's energy needs would be well over 50% higher in 2030 than today. China and India together account for 45% of the increase in demand. (IEA 2008)

There is a global challenge to shift to a more secure, lower-carbon energy system, without damaging economic and social development at large. IEA (2008) states 'to achieve a much bigger reduction in emissions would require immediate policy action and technological transformation on an unprecedented scale'. The potential to develop a system in space to harness clean energy from the sun and beam it back to Earth is without question a feat of 'technological transformation on an unprecedented scale'. At present, solar energy as a renewable and clean source of energy is being selected as an energy choice for more and more people. The use of solar energy in multiple forms is spreading, aided by technological advancements. However, there are problems inherent in the use of solar power as it is collected today, since the energy is diffused, intermittent and unreliable in many parts of the world. The concept of collecting solar energy in space and transmitting it to Earth in the form of microwaves avoids these problems as these are more intense, the sun shines 24 hours a day in space, and the energy collected could be delivered almost anywhere. The space solar power (SSP) concept (also known as space-based solar power (SBSP), was developed by Peter Glaser. It was first proposed in 1968 and is still considered to be a very promising source of renewable energy. Despite its benefits, however, it still remains in the research phase of development. The potential size of this market is predicted to be US$100 billion by 2020, and the European Space Agency, the Japanese Space Agency and the private Space Island Group are working together to commercialize this new source of energy (SPACE.com 2009).

Metals – ferrous, precious and strategic

Cobalt, gold, iron, magnesium, nickel, platinum and silver are raw metals that are becoming increasingly rare on Earth but can be found in abundance in space. The approximately 3,000 NEAs, 750 of which are easier to reach than the moon, contain invaluable quantities of these metals: 'The NEA Amun, about two kilometres in diameter, contains far more metal that the total amount used by the human race since the beginning of the Bronze Age. Its Earth-surface market value is tens of trillions of dollars, larger than the annual gross global product of Earth' (Fukushima 2008). If extracted, the different metals could be utilized for different uses, in space or on Earth. Some metals would be used to provide radiation shielding for explorers and colonists, while other ferrous metals could be employed in the building of space structures, and scientific samples or precious and strategic metals could be transported to Earth (Fukushima 2008). At present, mining specialists and space engineers are working on developing methods of extraction of these resources.

Fusion reactor fuels

Helium-3 is a potential energy source for fusion power reactors. It is predicted that the moon has about a million tons of this light helium isotope. 'Recovering a single tonne of helium-3 requires perfect extraction and recovery of all the gas from 100 million tonnes of regolith, a seemingly implausible amount.

Nonetheless, the energy content of the recovered helium-3 is so large that the process may still make economic sense' (Fukushima 2008). CNNmoney.com predicts that the value of one tonne of helium-3 could be worth US$7 billion, and that by 2018 the helium-3 mining industry on the moon alone could be valued at US$250bn. There is great interest in this fuel source from a number of governments and privately owned bodies, and it is anticipated that it will be actively mined by 2030 (CNNMoney.com 2009).

Military sector

As the number of countries with space programs continues to rise, so do government space budgets for both military and civilian applications. However, significant portions of military-related space budgets may not be revealed in published figures. Looking at public budgets related to space poses several methodological challenges. First, when they are available publicly in some detail, budgets may not necessarily match current expenditure. Second, published budgets may not reveal certain confidential segments of space programs (for example, for military purposes). Third, some expenditure may be classified under other areas of government outgoings – for example, telecommunications or R&D, rather than under 'space'. Finally, data were not available for all OECD member countries (though they were available for all major space participants). For example, data were not available for Australia, Iceland, Mexico, New Zealand, the Slovak Republic and Turkey (OECD 2004).

We have a more detailed picture of the space budget in the USA (see Figure 2.13), where the budget allocated to defense and military purposes fell sharply after 1990 to 44 percent, but has been rocketing since 1995 to reach 56 percent in 2007 (see Figure 2.14).

In contrast, the distribution between military and civil programs has been more stable over time in Europe (Eurospace/ASD 2008).

Space exploration sector

Space exploration is the physical exploration of outer-Earth objects, via robotic probes and human missions. More broadly,

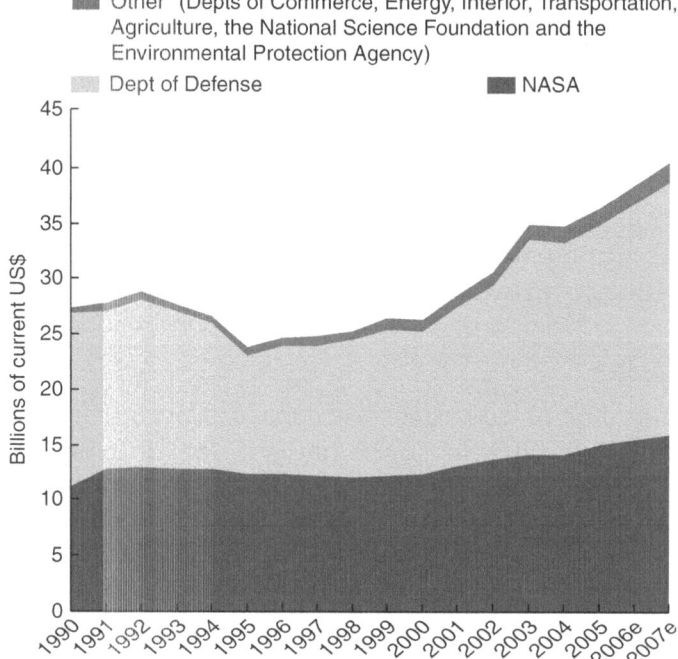

Source: OECD (2007).

Figure 2.13 US government total space budget, 1990–2007

it also includes the scientific disciplines (for example, astronomy, solar physics, astrophysics, planetary sciences), technologies and policies applied to space endeavors.

Countries with space programs are investing increasingly in 'down-to-earth' space applications (for example, telecommunications, earth observation), for strategic and economic reasons. Nevertheless, space exploration remains a key driver for investments in innovative R&D and the sciences, and constitutes an intensive activity for major space agencies.

Space exploration is probably the most visible face of space activities, constituting an inherent mission of space agencies worldwide. Its achievements generate enthusiasm among the public and wide media interest, as shown by the race to the moon, Mars exploration by robots and the probe landing on Titan, one of the moons of Saturn. Space sciences and planetary missions

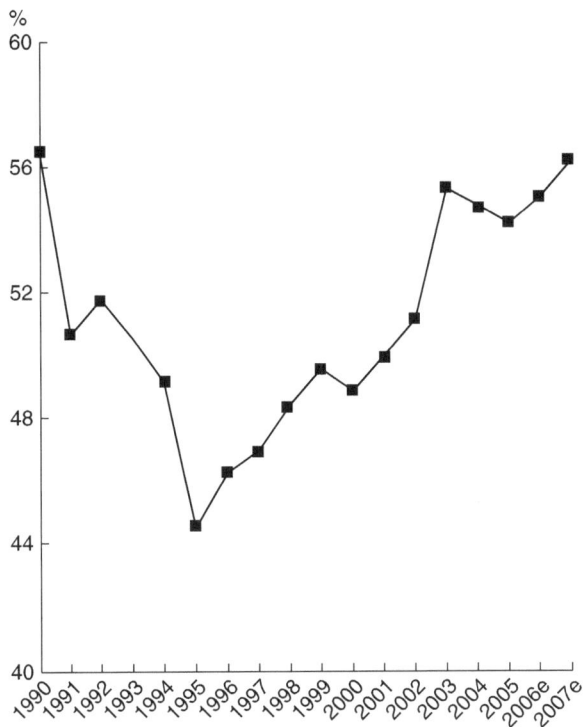

Source: OECD (2007).

Figure 2.14 Military expenditure as a percentage of the US space budget, 1990–2007

have developed markedly over the years. This trend is reflected in the current and planned robotic exploration missions of the solar system, in which the USA, Europe and several Asian countries are active players, as shown in Table 2.1.

Robotic missions presented here include active and planned orbiters (spacecraft whose purpose is to orbit a planet or an asteroid, usually to map its surface); planetary rovers (robots landing and roving on celestial bodies); and other exploration probes (spacecraft sent to fly by several celestial bodies). Planned missions may be cancelled, therefore only missions intended to be launched by 2008 have been included, according to the data available. Several dozen exploration probes have been launched over the years as national or international missions, targeting planets, moons, comets and asteroids in the solar system.

Table 2.1 Selected active and upcoming robotic exploratory probes, as of December 2006

Name of mission	Date of launch	Agency(ies)	Mission description
Lunar Reconnaissance Orbiter	2008	NASA (USA)	Lunar orbiter
Change 1 ('Moon Goddess')	2007	CAST (China)	Lunar orbiter
Chandrayaan 1 (Hindi for 'Moon Craft')	2007	ISRO (India)	Lunar orbiter
Selene	2007	JAXA, ISAS (Japan)	Lunar orbiter
Dawn	2007	NASA (USA)	Rendezvous and orbit asteroids Vesta (2011) and Ceres (2015)
Phoenix	2007	NASA (USA)	Lander to dig soil on northern plains of Mars and look for water-ice evidence (2008)
New Horizons	19 July 2006	NASA (USA)	On its way to Pluto and Kuiper belt (2015), flyby of Jupiter (2007)
Venus Express	9 November 2005	ESA (Europe)	Venus orbiter
Mars Reconnaissance Orbiter	12 August 2005	NASA (USA)	Mars orbiter
Messenger	2 August 2004	NASA (USA)	On its way to Mercury (2011), flyby of Venus (2007)
Rosetta 2	March 2004	ESA (Europe)	On its way to Comet Churyumov-Gerasimenko (2014), flybys of Asteroid 2867 Steins (2008)
Opportunity	7 July 2003	NASA (USA)	Mars rover

Spirit	10 June 2003	NASA (USA)	Mars rover
Hayabusa ('Peregrine Falcon')	9 May 2003	JAXA, ISAS (Japan)	Landed and collected surface samples from the asteroid Itokawa (2005); returned to Earth 2010
Mars Express	6 February 2003	ESA (Europe)	Mars orbiter
2001 Mars Odyssey	7 April 2001	NASA (USA)	Mars orbiter
Cassini	15 October 1997	NASA, ESA, ASI (USA, Europe, Italy)	Saturn orbiter (the Huygens probe carried onboard landed on Titan in 2005)
Ulysses	6 October 1990	NASA (USA)	Solar orbiter
Voyager 2	20 August 1977	NASA (USA)	Exploration outside the solar system (currently +12 billion kilometres away from the Sun)
Voyager 1	5 September 1977	NASA (USA)	Exploration outside the solar system (currently +15 billion kilometres away from the Sun)

Notes: In addition to those robotic exploration missions targeted at extraterrestrial bodies, more than a dozen space science satellites are in Earth orbit. Two large international space telescopes (NASA/ESA) were active as of December 2006: the Hubble Space Telescope (launched in 1990) and SOHO, the Solar and Heliospheric Observatory (launched in 1995). Hubble's successor, the James Webb Space Telescope could be launched in 2013. The international CoRoT observatory, led by the French Space Agency (CNES) (launched in 2006), and NASA's Kepler observatory (launched in 2009) are designed in particular to search for Earth-like planets outside the solar system.

Source: OECD (2007).

Manned space flight sector

In the case of human spaceflight, several definitions for 'astronaut' co-exist. The International Aeronautic Federation (IAF) calls anyone who has flown at an altitude of 100 km an 'astronaut'. The US Air Force set the limit at 50 miles altitude (80.45 km), while other organizations consider that a person must have reached orbital velocity and remain in orbit (above 200 km) to be considered an 'astronaut'. The IAF definition has been used here as it was used by the OECD (OECD 2007).

In addition to robotic exploration, the development of a human presence in space has been a recurring theme since the 1950s for both political and prestige-related reasons. Currently only three countries – Russia, the USA and China – have the autonomous capability to launch human beings into space; however, a total of 451 people from 37 different countries have flown in orbit around the Earth as of late December 2006 (OECD 2007) (see Table 2.2). Since the late 1990s, the feasibility of commercial human space-flight endeavors is also being tested via 'space tourism' ventures.

Table 2.2   Selected human space-flight statistics, as of December 2006

Countries with autonomous capability to launch humans into space	3[1]
Number of launches with humans on board	+240
Persons who have flown into orbit	451
Persons who have flown over the 100 km altitude threshold (including suborbital flights)	454
Number of nationalities who have flown in space	37
Astronauts who walked on the moon (1969–72)	12
Operational and inhabited space stations since the 1960s	9[2]
Professional astronauts currently in orbit (the International Space Station has been manned continuously since 2003)	3
Number of paying orbital spaceflight participants ('space tourism')	4

Notes:
[1] China, Russia, USA.
[2] 7 Russian, 1 US, 1 international.

Source: OECD (2007).

The current state of the downstream segment

Space-related services use a specific satellite capacity, such as bandwidth or imagery, as inputs to provide a more global service to business, government or retail consumers. Those services are as diverse as the space applications themselves. The services are traditionally divided into three large application domains: telecommunications; earth observation (also known as remote sensing); and navigation. Value chains often involve public agencies as investors and final users. As such, public authorities remain significant customers even in well-established commercial markets, such as telecommunications.

Space-related services revenues are not easy to gauge nationally or internationally, but worldwide estimates ranged from some US$60 billion to over US$100 billion in revenues in 2006 (see Figure 2.15).

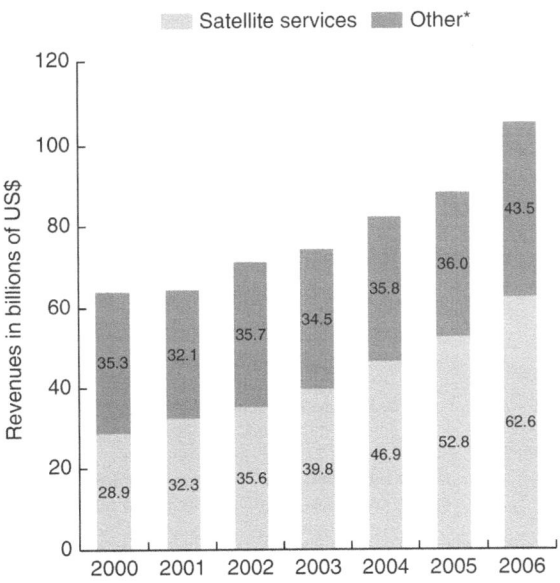

Note: *Ground equipment, launch industry and satellite manufacturing

Source: OECD (2007).

Figure 2.15 World satellite industry revenues for services and others

Satellite communications services

According to the US Satellite Industry Association (SIA), revenues from the world satellite services industry (mainly telecommunications and earth observation services) were 83 percent higher in 2005 than five years earlier, and still growing in 2006. Telecommunications services, in particular Direct Broadcast Satellite (DBS) services (for example, satellite television), represent the bulk of commercial revenues, with US$48.5 billion in 2006 (see Figure 2.16)

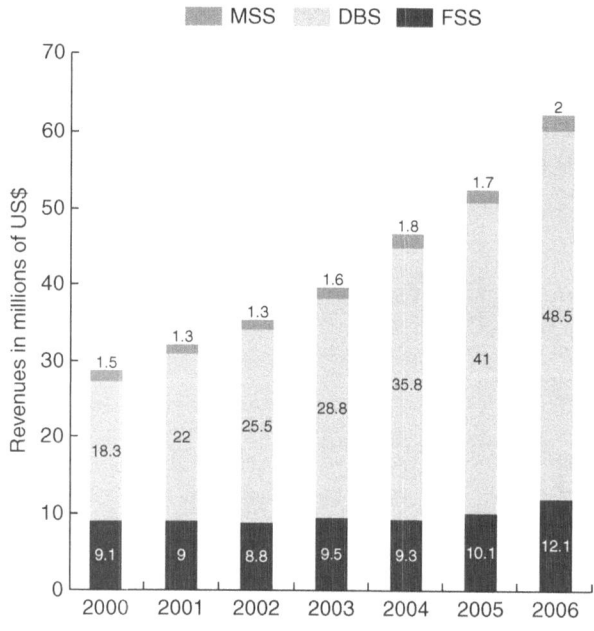

Notes:
MSS (Mobile Satellite Services): Mobile telephone and mobile data;
DBS (Direct Broadcast Satellite): Direct to home television (DTH),
Digital Audio Radio Service (DARS), and broadband;
FSS (Fixed Satellite Services): Very Small Aperture Terminal
(VSAT) services, remote sensing, and transponders agreements.

Source: OECD (2007).

Figure 2.16 World satellite service revenues

The biggest and the most mature downstream segment of the space sector is the telecommunications segment. It has two major parts: telecommunications and broadcasting, we can also differentiate between mobile and fixed services. The digital revolution has been the major development in recent years. This revolution has resulted in the diverse services converging, and an increasing demand for sophisticated services with expanding requirements in terms of bandwidth.

Further growth is expected as a result of expected satellite operators' consolidations and strong demand worldwide. Other space-related services, in Earth observation and navigation, are not generating as much revenue (see Figure 2.17), though governments, particularly defense departments, increasingly use satellite capacities, as demonstrated by their use of commercial satellite bandwidth (see Figure 2.18).

Concerning Earth observation, satellite imagery should benefit from increasing worldwide demand for geospatial products (for example, weather forecasting) (see Figure 2.19).

The satellite operator industry

The satellite operator industry can be further segregated into three service types, namely broadcasting satellite services (mainly TV and radio); fixed satellite services (operating, servicing and leasing satellite capacity); and mobile satellite services (including voice traffic), bringing total revenues to US$62.6bn in 2006. The great majority of this expenditure, US$48.5bn (77.5 percent), can be attributed to the broadcasting satellite services, US$12.1bn (19.4 percent) to the fixed satellite services and the remaining US$2bn (3.1 percent) to mobile satellite services (Futron Corporation 2009).

As pointed out above, broadcasting services are divided into satellite television and satellite radio services. Satellite television is by far the dominant component, with revenues of US$46.9bn (97 percent) compared to US$1.6bn (3 percent) in revenues from satellite radio (Futron Corporation 2009). Fixed satellite services are divided into a broad number of categories, leasing transponder capacity being the most important one. Mobile satellite services can be divided into two groups: mobile data and mobile voice services.

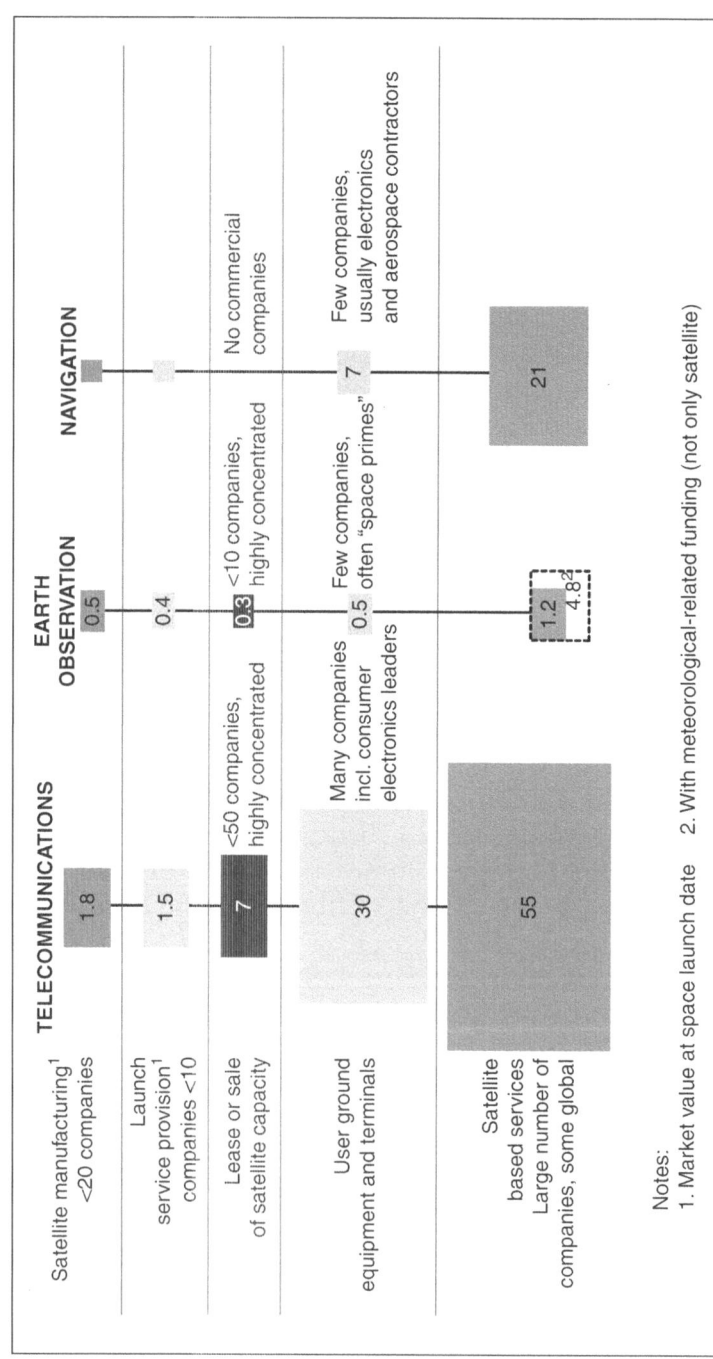

Source: OECD (2007).

Figure 2.17 The three value chains in commercial satellite applications, 2005, in billions of US$

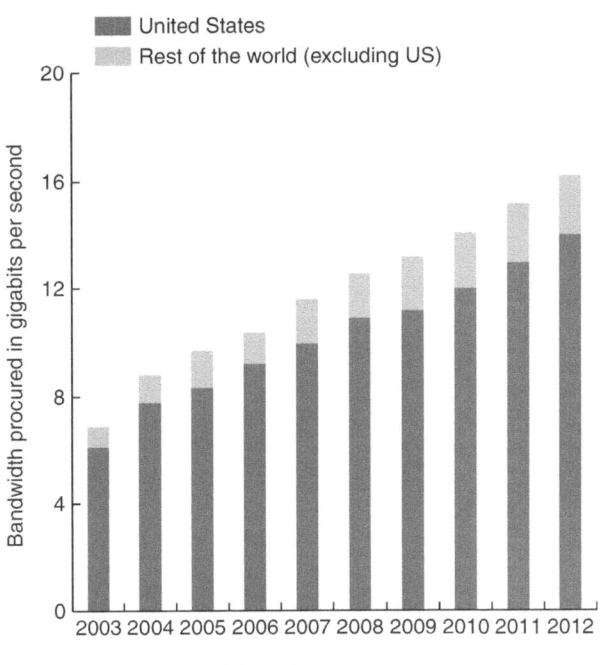

Note: Estimated 2007–12.

Source: OECD (2007).

Figure 2.18 World government and military commercial satellite market total, 2003–12

Satellite subscription and retail services – the direct-to-home industry

According to a recent report from the Futron Corporation (2009), the direct-to-home (DTH) industry is the largest section of the industry and its development is quite recent. France, for example, adopted its first analogue DTH platform in 1985; and the world's first digital DTH platform, Orbit, began operations in 1994. The DTH operators have derived benefit from the technological progress achieved upstream in satellite production and downstream in digital compression in addition to market liberalization. From 1995 to 2001 growth was intensely rapid. The Direct Broadcast Satellite (DBS) industry revenues increased from US$1.5 billion to US$22.5 billion; in 2001, fifty-four DTH platforms broadcast more than 5,000 TV channels to more than 45 million subscribers all over the world. The main market

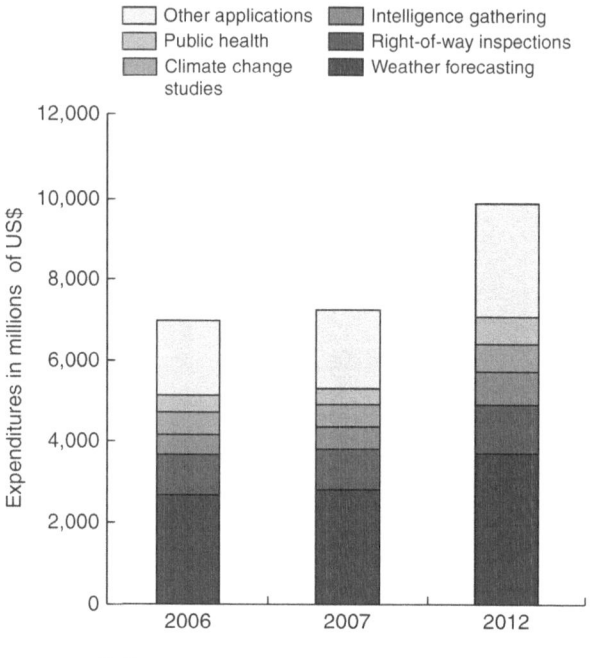

Notes:
1. Includes more than just satellite imagery, i.e. aerial.
2. Estimated 2007, 2012.

Source: OECD (2007).

Figure 2.19 Estimated global expenditure for remote sensing products, by application, 2006–12

segments are movies (23 percent), documentaries (12 percent), sport (10 percent) and news (8 percent) and the revenues of the DTH platforms exceeded those of the box office and video games in the years 2003 and 2004. The great success of DTH is largely a result of progress in the productivity of DTH satellites since the late 1990s, which is again a consequence of an increase in satellite durability, in the number of transponders per satellite, and in the population covered. The revenues of the fifty-four DBS operators in 2003 amount to US$33 billion, an upward movement of 27 percent compared with the previous year (OECD 2004). More recent data indicate that DTH television terminals account for 84 percent of all terminals in service. In comparison, terminals to receive internet broadband services account for only 0.5 percent of all terminals in service (Futron Corporation 2009). It is obvious that the necessary ground equipment is a prerequisite for using satellite signals, and an increase in demand for satellite services such as broadcasting will drive up the investment in hardware for such items as satellite dishes, thus opening up opportunities for future price reductions because of the scale.

Video services such as television channels have seen strong growth in recent years. There are various explanations for this. The expansion of high-definition (HD) content (HDTV) has attracted many new users to satellite technology as a convenient way of exploiting the new technology. More mature markets such as those in North America and Europe will derive growth from this technological breakthrough. According to the data available (Futron Corporation, 2009), HDTV channels, which account for only 2 percent of all channels carried through satellite, are expected to see staggering growth, with annual expansion rates of up to 35 percent between 2010 and 2015.

Also, the fact that many developing markets are able to receive satellite television will drive the demand from broadcasters for satellite video services. Increasing wealth in emerging economies will increase the audience that can afford these services. For example, India has seen its market opening up and licenses have been given to transmit satellite signals directly to house-holds via DTH technology. This gives customers more choice and will drive up demand for video services as channel variety increases. On the other hand, some markets with high growth potential, such as China, are still facing restrictions. As a result, DTH content providers of video services will lose out in those markets. DTH is also experiencing a late boom in the US market at the time of writing, despite its being introduced as long ago as the late 1980s. Nevertheless, it is forecast that future growth will come increasingly from emerging markets, which are currently still lagging behind the American and European markets (Futron Corporation, 2009).

Interactive broadband services

Enhanced and interactive broadband services that can be included in traditional DTH transmission services may be the driver for the decade 2010–20. The clear advantage of satellite broadband over terrestrial technologies is that satellite broadband can provide worldwide coverage more or less immediately with minimal development of infrastructure. On the other hand, it has notable disadvantages: the service is not reliable at all times, and costs tend to be very high. Furthermore, because of latency (that is, the time it takes for a piece of information to cross a network connection, from sender to receiver), satellite broadband is not appropriate for services requiring a high level of interactivity, such as gaming. Thus satellite broadband is a temporary arrangement for countries with a fairly underdeveloped terrestrial technology, and it can be sold as a niche product for rural and remote areas in countries with a rich terrestrial offering (OECD 2004).

Forecasts for growth perspectives in the data markets are becoming more complex, since many services such as voice transmission are increasingly being transferred through data networks such as VoIP (Voice over Internet Protocol). The major growth in the data sector will most likely be related to overcoming the digital divide in terms of the internet and other forms of communication. Regions or countries that did not have the capital or ability to connect to the conventional broadband networks will now be able to receive satellite services as the satellite covers the region in any event, and only ground equipment will be required to start receiving the signal. Governments have a substantial interest in benefiting from such developments, and data markets will in turn be able to benefit from increasing demand for their services.

According to the report published by the Futron Corporation (2009), the highest overall growth in demand for data services is expected to come from Asia, Latin America, the Middle East, certain parts of Africa, and Eastern European countries. The introduction of advanced technology has contributed greatly to improving the price performance of these fixed satellite services in recent years. Because of the reduction of the number of analogue channels transmitted, satellites in space now have spare capacity to push for augmenting the data services extensively.

Growth in voice services over satellite will be limited to regions that are actively expanding satellite telephone services in countries in sub-Saharan Africa (excluding South Africa). VoIP will increasingly take over and, as a result, growth opportunities will remain limited in this small market.

To summarize, the growth opportunities for media content over satellite, and DTH in particular, will continue to provide the satellite industry with numerous and diverse opportunities. The largest portion of that growth will come from the video services that will remain the largest market for satellite capacity for the decade to come. Data services will also see growth while voice services will be forced to face substitution.

Global positioning and navigation services

Only the US Global Positioning System (GPS) is fully operational at the time of writing, but the use of satellites for location and navigation purposes is increasing rapidly. GPS comprises a network of twenty-four satellites launched in the medium Earth orbit (20,200 km) and was originally developed by the US Department of Defense to maintain troop contact and be aware of troop positions. GPS is at present a standard feature in everything from search and rescue services to automobile navigation and leisure goods. GPS created a significant downstream market estimated at about US$10.6 billion in 2001 (including hardware and services), and it is projected to reach US$28.9 billion by 2015, according to a new report by Global Industry Analysts (Global Industry Analysts Inc. 2010).

GPS equipment and chipsets drove this growth, increasing by 43 percent to US$40.7 billion in 2006 to become one of the largest revenue producers in the space industry (Eurospace 2008). However, as the use of the US GPS becomes omnipresent, and as an increasing number of systems depend on it, there is growing concern that a disruption in the signal could have tremendous global consequences. Hence several countries are determined to develop their own positioning and navigation systems, with the European civil Galileo system expected to both complement and compete with GPS.

The Galileo system is a dedicated global navigation satellite system that has been conceived by the European Union (EU) and the European Space Agency (ESA). Galileo should be operational by 2013. The system will provide more accurate measurements than the ones available through the US GPS or the Russian GLONASS, improved positioning services, and a sovereign positioning system upon which European countries can rely in situations of political disagreement or outright war (BBC 2007).

Earth observation

Commercial observation satellites (COS) are relatively new, though Earth observation was one of the earliest applications of satellites. When restrictions on satellite imagery technologies were eased at the end of the Cold War, the industry began to grow. Technology has played a central role in the improvement of COS, particularly progress in optical and radar sensor technologies, enabling the development of smaller satellites that were cheaper and more mobile than the early COS. However, the expected returns from COS are uncertain. On the one hand there is the perception that a wide range of countries will shift to knowledge-based economies, and so a fast-growing market for satellite imagery and related information products may emerge globally over the next few years. On the other hand, however, COS are confronted by stiff competition for selling geospatial information products (OECD 2004). For example, COS have to compete with aerial photography and land-based surveys using global navigation satellite systems (GNSS) and geographic information systems (GIS) that both compete with and complement COS imagery (OECD 2004).

Space tourism

When Dennis Tito became the world's first space tourist in 2001, despite initial reluctance from his own national space agency, NASA, he proved it was possible and realistic for an ordinary individual to expand his/her horizons beyond this planet as a 'humble tourist'. It may have been a breakthrough for extreme tourists, but more importantly, the business world sat up and took notice, for the vast potential for commercial exploitation suddenly became far more realistic and ultimately, attainable. Tito's foray into space has essentially created a whole new space race, where a scramble to take advantage of space tourism's massive profit potential is now well under way.

The idea of space tourism is not actually a wholly fresh one. In fact, while riding on the crest of a wave after the first moon landings, the first serious and detailed considerations for space travel were mooted in the 1970s. However, difficulties were greater than expected and the infamous Challenger disaster in 1986 forced a rethink, with all notions being shelved indefinitely. However, rapid technological advancement throughout the 1990s reignited interest in space tourism, and it is now a realistic possibility once more. In addition, a wealth of studies on space tourism has been carried out in such varied countries as Germany, the UK and Japan. The wealth (and breadth) of renewed interest and engineering research has greatly accelerated the movement, particularly during the 2000s. One of the biggest breakthroughs was made in 2004. In the quest to find a suitable vessel to ferry prospective customers, the US$10 million Ansari X PRIZE was offered as an incentive for a non-government organization to launch a reusable manned spacecraft into space (X PRIZE Foundation 2009). The competition was an undoubted success, with Scaled Composites' SpaceShipOne successfully completing the groundbreaking voyage, and ultimately opening up an entire new realm of possibilities. SpaceShipOne aims to provide sub-orbital trips – those that just breach the atmosphere. On the other hand, orbital trips are regarded as the 'real deal', in that the passenger actually goes into orbit, as Tito did. With this remarkable progress taking place, more countries are now getting involved, breaking the traditional US–Russia duopoly and ensuring that space is now becoming a more global industry. Remarkable investments and interest has duly followed in order to capitalize on what could now potentially become one of the twenty-first-century's most lucrative industries (Eurospace 2008).

There is a wide array of companies working on a breakthrough and vying to become the key player in the field. In an emerging and competitive market, probably the most renowned player thus far is Virgin Galactic, another of Richard Branson's Virgin offshoots. Given Branson's world-renowned penchant for prolific success from the most eclectic of projects (Branson 2006), there must be substance behind the belief that this industry is potentially lucrative. Virgin Galactic has been determined to make many of the initial inroads by seeking early competitive advantages. In the immediate aftermath of SpaceShipOne's historic flight, Branson saw his opportunity and moved quickly to purchase Scaled Composites, which is now part of a joint venture with Virgin Galactic (Branson 2006). Arguably the largest player in the industry to date, the group is currently enhancing the space plane that will become known as SpaceShipTwo. With this craft, Virgin Galactic announced on 10 October 2010 the successful completion of the first piloted free flight of SpaceShipTwo, named the VSS *Enterprise*. Virgin Galactic is now well on the way to becoming the world's first commercial space line, with 370 customer deposits totaling US$50 million. Future commercial operations will be at Spaceport America in New Mexico inaugurated on 22 October 2010. Virgin Galactic is striving to establishing itself quickly as the forerunner, but there are rivals aplenty. Others are also in the process of developing reusable spacecraft, ensuring that Branson will have stern competition. The key player that has gained the most progress so far, however, is Space Adventures. Apart from offering exclusive orbital trips as opposed to sub-orbital (through Soyuz taxi missions), it was Space Adventures who were the providers behind Tito's trip, and have gone on to provide subsequent trips to four more 'tourists', with all the trips that have operated so far being sold for a total of a whopping US$120 million (Space Adventures 2009).

Of course, it is quite obvious that, at present, the average individual cannot afford this ultimate luxury trip (Goehlich 2005). All of those who have travelled to date with Space Adventures have been multi-millionaires whose bumper disposable income enabled them to splash out on this, the most original of holiday destinations. It is at present, an extremely selective market, with Tito and fellow pioneer Mark Shuttleworth having paid out approximately US$20 million each for the privilege of their trip.

In an extensive report entitled 'Suborbital Space Tourism Demand Revisited', Futron Corporation have explored the market outcomes in more detail by conducting some extensive market research (Futron Corporation 2009). With price ranges being primarily taken into account, Futron, in collaboration with Zogby Polling Firms, surveyed 450 wealthy people (as defined by either an annual salary of at least US$250,000 or a net worth of US$1 million) to investigate the potential demand.

The survey also took into account key customer considerations, including downsides, potential risks and dangers, in order to depict an accurate image of space tourism. Futron reveal that passenger demand was actually down from 15,000 passengers to 13,000 compared to their original survey carried out in 2002 (Futron Corporation 2009).

In addition, Virgin Galactic announced that they are selling seats for their initial flights at US$200,000, double the price suggested in the original survey (Futron Corporation 2009). These core factors have obviously profoundly influenced the outcome of the market predictions, and as a result projected potential revenues in 2021 have decreased from US$785 million to US$676 million (Futron Corporation 2009).

However, Futron also pointed out the dynamic nature of current demand and price settings when illustrating that 'the impact of initial ticket prices actually produces higher revenues in some parts of the forecast period, reflecting the interaction of price changes and user demand'. As foreseen earlier, it is mainly the gradual decrease in price settings that will prove crucial to the progress of this industry, and pave the way for expansion into more market segments, perhaps by moving from marketing exclusively to tycoons to include those with a lower income.

The issue of price has been a predominant feature for obvious reasons, but it is important to note that this may not be the only stumbling block for potential customers and prospective companies (Billings 2006). Other factors have emerged in the development of this industry, such as customer considerations, legislation and government interventions. In terms of customer considerations, it should be noted that a sub-orbital flight (let alone an orbital flight) is not an average, slightly gruelling long-haul flight. The incentives are spelt out clearly, such as customers experiencing what up to the present only astronauts have seen, the acceleration of a rocket launch and going 50 miles into space (Goehlich 2005). It is also an inherently risky activity.

Apart from the vehicle having a very limited flight history, it would also be imperative to undergo training prior to the trip. Burt Rutan, the aerospace designer behind SpaceShipOne and chief of Scaled Composites, acknowledges that extensive improvements in safety must be prioritized over improvements in affordability for the present: 'We believe a proper goal for safety is the record that was achieved during the first five years of commercial scheduled airline service which, while exposing the passenger to high risks by today's standards, was more than 100 times as safe as government manned space flight' (Rutan 2006). Rutan knows how pivotal the safety and customer consideration elements are. If safety goals are realized and prices driven down, this market could expand spectacularly: 'However, with ticket prices well below $50,000, it is believed that there could be the order of 500,000 space trip passengers per year' (Rutan 2006).

Such talk of safety does not convince the skeptics, and the US Government in particular. Cynics may scoff and declare that NASA has felt threatened by the sudden evolution of space as a mere business entity, as opposed to an infinite realm of research. Its apprehension toward space tourism has been well documented(Collins 2002; Loizou 2006; Billings 2006). Consequently, all firms looking to cash in on space tourism cannot ignore the influence of this well-established government agency. It looks as though they will need to cooperate heavily with NASA in the next crucial years of development. NASA was a vehement opponent of space tourism until the International Space Station (IIS) was completed in 2006. Interestingly, though, NASA's European counterpart and competitor, the European Aeronautic Defence and Space Company (EADS) (which has increasingly narrowed the gap with the US industry) have appeared to embrace the challenge of development as a new business opportunity. As a result, they are relishing the challenge of providing holidays in orbit by developing their own sub-orbital craft, and to emerge as yet another key global player in the industry: 'EADS Astrium has a lot of expertise when it comes to operating in space, and there is little doubt that it has the engineering resources needed for such a project' (Cunningham 2007).

Apart from NASA's apparent reluctance, legislation also seems to be an imposing obstacle that will need to be overcome. A multitude of legal and regulatory aspects of public space travel and tourism must be resolved before any viable large-scale businesses can truly emerge and begin to operate regularly.

To illustrate their breadth, this legislation includes, to name but a few items: 'identification of public policies and/or laws that exist or must be enacted to enable business formation, licensing, certification and approval processes for both passengers and vehicles, clearance and over-flight considerations, and environmental and safety issues including atmospheric pollution, solar radiation and orbital debris' (O'Neil 2009).

There is far more room to maneuver in the area of infrastructure. The development of SpaceShipOne has obviously been the key development to date, with Scaled Composites quickly turning their Mojave base in California into an epicenter for space tourism. Expansion is already occurring too, with another site in the pipeline (likely to be situated in New Mexico) to provide another launch base and significantly improve the infrastructure of the industry. Such progress is already causing positive knock-on effects with the signs of 'space fever' slowly becoming more transparent: 'A variety of ground facilities and activities may be called into play to support successful formation of general public space travel and tourism enterprises, especially including businesses that are formed to exploit the latent market demand even before actual in-space trips are available to the public' (O'Neil 2009).

The impending reality of the 'space gold rush' has spurred more aspects of the industry to sprout, and in terms of infrastructure, accommodation has also now become the newest feature. Plans for space hotels are now in various stages of development, with new companies such as Bigelow Aerospace and the Space Island Group making early headway. Even Hilton Hotels has made plans in this direction, illustrating that the industry is very real, and that commercialization is officially in full swing (Cunningham 2007).

3

The main country players

Introduction

At the time of writing, space exploration and space technology are paid considerable attention in many governmental policies across the globe and are also attracting much attention from the private sector. Nations such as the USA, Russia, Japan, China, Brazil and Canada, as well as collaborative space agencies such as the European Space Agency (ESA) in Europe have formulated and published their visions for the future of the space industry. So too have an increasing number of private ventures through Burt Rutan, Richard Branson and others.

Currently, much of the vision surrounding the next generation of space missions and technology is tied to the perceived 'second race to the moon' and beyond. This civilian theme is complemented by an ongoing discussion about the military facets of space activity, as well as the role of both current and emerging commercial enterprises in space access and exploration. Together, the civilian, military and commercial space sectors focus the broader space discourse on questions about the elements of space competitiveness, the relative competitive position of traditional space leaders, and the role of emerging space powers such as China and India.

When we examine the literature published by the different bodies involved in this industry, some common themes are easily identifiable. NASA's strategic plan, published in 2003, sums up these themes most succinctly with its maxim: 'To improve life here, to extend life to there, to find life beyond'. NASA, of course, pioneered human entry into space alongside the Russians in the third quarter of the twentieth century. The world's two super-powers at the time were engaged in the 'space race', some might argue, as a game of one-upmanship during a period when relations between the two ideologies of capitalism and communism were in direct competition to overpower each other.

Today, however, there is a shift toward collaboration between the nations of the world to advance human knowledge, and harness the potential benefits that space presents. Former Russian president and now Russia's prime minister, Vladimir Putin, acknowledged recently that: 'for the modern ... world nations, cosmonautics now is not only the subject of national pride. Exploration and application of Earth-orbital space become serious resources of national development and real advancement of people's living standards'.

The present-day space industry has evolved from the romanticism of the 1960s and 1970s, when putting a human being on the moon captured the imagination of the world (though it should be noted that until the change in strategy announced recently by the Obama Administration, NASA had credible intentions of returning, permanently, to the moon by 2020). Now, a multi-pronged approach to space exploration is attempting to address environmental issues, advance technology and industry, and cater for the next generation of holidaymakers – the space tourists.

A number of factors have contributed to the globalization of the space industry. Political changes in the 1990s and the end of the 'space race' meant that almost all trading nations function with market-based economies and their trade policies have tended to encourage free markets between nations. The globalization of the space industry has been further encouraged by technical standardization between countries. Host governments actively seek to encourage global operators to base themselves in their countries (namely, the US space infrastructure, Russian know-how, Brazilian lower launch costs). The national home base of an organization in the space industry plays an important part in creating advantage on a global scale. As elaborated in Porter's Diamond of National Advantage (Porter, 1998), the determinants of national comparative advantage stem from demand and factor conditions, firm strategy structure and rivalry, and related and supporting industries. In countries such as the USA and Russia, experience in space technology and space infrastructure provided initial advantages that have been built on subsequently to yield more advanced factors of competition. At the same time, home demand conditions (American early adopters in the first stages of space industry segments' life cycles), technology transfer, related and supporting industries (American, Russian, Japanese technological supremacy) and domestic rivalry have also provided a basis upon which the characteristics of the advantage of an organization competing at the new frontier of international competition have been shaped. Yip (2003) cites decreasing costs, globalization, scale economies, sourcing efficiencies as offering the potential for competitive advantage to some countries. Table 3.1 provides a cross-country SWOT analysis.

A recent report from the Futron Corporation (2009) addresses strategic pivotal questions about space power and competitiveness:

- What are the core measures of space competition?
- Is 'space nationalism' on the rise, and if so, what are the implications?
- What is the current positioning of traditional space powers such as the USA, Europe and Russia?
- What role will emerging powers such as India and China play? Partners or competitors?
- What is the competitive role for lower-tier players such as Japan, Brazil, Israel and others?
- What are the implications of a multi-polar space community?
- What are the economic consequences of a commercial space environment based on multiple international providers of key technologies, systems and services?

To provide practical insight into these strategic questions, Futron published the first annual Space Competitiveness Index (SCI),

Table 3.1 Cross-country SWOT analysis

	Strengths	Weaknesses	Opportunities	Threats
USA	Dominant world player in space activities. By far the biggest budget of all nations involved in space, with US$15.5bn in 2004. Already sent missions to Mars while most others are still planning to orbit the moon (Mustafa 2004).	Has come in for strong criticism over its role in the International Space Station – namely that time and money could have been better spent on other space projects.	Intention to return to the moon by 2020 (though now apparently stalled by the Obama Administration) and developing a permanently manned lunar outpost.	Growing power of all other players could diminish America's dominance.
ESA	Fifteen-nation collaborative effort. Second-largest annual budget of all space agencies – US$3.6bn.	So far has been reliant on collaboration with USA to get into space.	Future collaboration with others, such as Russia, likely. Intention of missions to Mercury and to land on a comet by 2014.	Other economies (China, India, Russia, etc.) are growing faster than Europe and hence so are their budgets, causing ESA to lose its advantage over them.

(Continued)

Table 3.1 (Continued)

	Strengths	Weaknesses	Opportunities	Threats
Russia	Long history and knowledge base built up during Cold War and space race with USA.	Space program stagnated and fell years behind USA because of break-up of the Soviet Union and loss of space race. Severe budget problems.	Seems likely to partner with Europe in future attempts at manned space flight.	Budget problems and distance from the USA technically compared to where it used be mean it will struggle to be a world power in space again.
China	One of only three counties to have twice sent astronauts into orbit independently.	Space program more about national pride and defence than scientific advancement (Fairclough 2007).	(i) Plans to put a man on the moon between 2020 and 2024 (ii) Collaboration with Russia.	Looming space race with Japan to put a man on the moon could be costly – financially and to national pride.
Brazil	Operates its own independent launch site at Alcantara. It is near the Equator so launch costs are lower and safety conditions are increased relative to others.	Still a burgeoning force in space matters. Only had its first sub-orbital rocket launch in 2004, following a failed launch in 2003 that killed 21 technicians.	Having previously relied solely on the USA for space programs, it has branched out and begun working with others such as China, Argentina and Israel.	The launch site at Alcantara and the VLS rocket are under the military's jurisdiction and not the space agency's, leading to confusion over the space agency's position as head of the space sector in Brazil (Silva 2005).

Japan	While behind China chronologically, its program is seen as being more complex and advanced (Hess and Kossack 1981).	Has not developed its own manned spacecraft yet despite its ambitious and fast-paced goals for the future.	Goal of landing on the moon to complete building of a lunar base by 2030.	Has shifted away from international efforts since 2005, concentrating on independent space missions.
UK	Member of ESA and has its own independent space centre.	Has typically been opposed to manned space flight in favor of robotic ventures.	Sees itself as a world leader in Earth observation, navigation and positioning.	Falling numbers completing space science and engineering degrees in UK will cause skills shortage in the future.
India	Has more than survived as a space player on a 'shoestring' having been held at arm's length by the others (Morring and Neelam 2004).	Largely indigenous industry concerned more about itself than collaboration or the needs of others it could satisfy.	Largely civilian aspirations involving communications, agriculture and meteorology.	Limited collaboration on space projects.

which systematically analyzes, evaluates and ranks the top ten global space participant countries across some forty metrics representing the underlying economic determinants of competitiveness within three major dimensions: government, human capital and industry. According to the most recent ranking published (Futron Corporation 2008), the countries that are better endowed according to these three main dimensions jointly considered were namely the USA, Europe, Russia and China. At the lower end of the ranking, as the less well endowed countries, we find countries such as South Korea, Israel and Brazil.

In relation to the ability of governments to provide structure, guidance and funding, the report states that transparency regarding space strategy, policy and spending remains a significant issue with some countries, particularly in connection with military space activity. This reduces the ability of commercial space actors to optimize investment and participation in the industry.

The USA has the most robust government policy-making structure (scoring 39.43 out of 50), including detailed strategies for military, civilian and commercial applications. The United States also spends more money – both military and civilian – on space capabilities, though investment is skewed toward military budgets. Other countries that performed well under this dimension were Europe (18.23/50), Russia (16.41/50), India (10.58/50) and China (10.44/50). At the lower end of the ranking as the less well endowed are countries such as Israel (6.13/50), South Korea (6.09/50), and Brazil (4.01/50).

European governments, through the European Union (EU) and ESA, have well-developed policy-making structures and are increasing spending, particularly within the civilian and commercial spheres. Japan recently reorganized its space agency and is considering significant updates to its space laws.

As for the ability for people to develop and use space applications and technology (that is, human capital), the Futron Corporation's report relies on the human capital scores by country The availability and development of sufficient human resources for the space sector, particularly in technically skilled areas such as engineering, is a concern for most countries, while data regarding human resources within the space industry is sparse, and lacks consistency across countries. In particular, countries that performed well under this dimension were namely USA (12.85 out of 20) Europe (9.02/20) and Russia (5.59/20). At the lower end of the ranking as the less well endowed we find countries such as South Korea (1.83/20), Israel (0.90/20) and Brazil (0.18/20).

In relation to the ability of the industry to finance and deliver space products and services, the Futron Corporation report concludes that satellite communications is the one market segment that is principally in the hands of the private sector. There is a significant increase of commercial interest in Earth observation, as well as a rapidly growing downstream market based on the US-operated Global Positioning System (GPS). The US commercial space industry is the clear leader (ranking 39.15 out of 50), followed by Europe (20.82/50) and Russia (12.86/50). While both Russia and China (4.33/50) score in the top five, government-controlled entities dominate their space industries, though China has proposed the stimulation of commercial space activity as a strategic objective. Likewise, the state dominates in the case of Israeli space activities (1.35/50), with some private sector ventures emerging.

According to the Futron report (Futron Corporation, 2008), based on the three dimensions (government support, human capital, and industry), the USA is still the current leader in space competitiveness, followed by Russia, Europe and China.

Similar evidence to assess the space power of a given country over time can be inferred by looking at the number of orbital satellite launches.

While from 1998 to 2007 Russia retained the top position in terms of number of launches, with the USA in second position, China has recently surpassed Europe in terms of the number of launches per year (Futron Corporation 2008).

By also looking at satellites manufactured by countries between 1998 and 2007, it also emerges that US satellite production has recovered somewhat from a steep fall in the early 2000s, but output still remains well below that of the 1990s (Futron Corporation 2008).

As for satellite manufacturing, over the ten-year period 1998–2007, the USA accounted for just over half of all satellite manufacturing (with 52 percent of the world's production), with European and Russian output accounting for just over a third (respectively, with 15 percent and 20 percent). The other satellite manufacturing countries represent relatively small volumes of output.

USA

The potential for commercial opportunities in the use of space by selling goods and services to governments and private customers is increasingly apparent (Hertzfeld 2007). While the USA has been the technological and commercial world leader in space, exercising 'space power' or control of outer space since the 1950s, competition is growing. At the time of writing, companies in the USA are in direct competition with many foreign entities in virtually all areas of the business of space, extending from launch vehicles, remote-sensing satellites and telecommunications satellites of all kinds (voice, direct TV, fixed and mobile services) to navigation services. Not only is there an increasing diversity of players active in the business of space but, as Hertzfeld (2007) has noted, the unique advantages of the space environment have also contributed greatly to the growing trend toward globalization through its almost universal coverage of populated areas with communication and observation products and services. Expanding markets and greater access to markets as a result of the globalization process has further driven the growth of the business of space (Mustafa 2004).

Because of the strategic value of space (see Figures 3.1 and 3.2) as well as the huge dependence of almost every industry on the space infrastructure, space commands a special importance and has become a critical national resource.

Hertzfeld (2007) has described how space power can be viewed from a commercial perspective in two ways. The first is

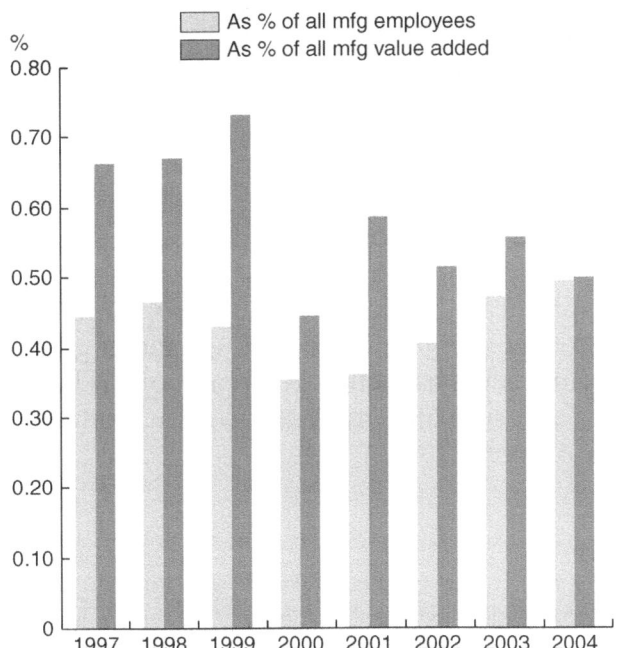

Source: OECD (2007).

Figure 3.1 Contribution of the space industry to the US economy, 1997–2004

economic: encouragement of US commercial space ventures to be dominant in the world marketplace, either through the creation of a monopoly or by sheer market dominance. The second is by aggressively denying others access or interfering with the operations of foreign space assets. Hertzfeld concludes that it will not be possible for commercial interests to supersede other national interests in space. Since nations are responsible for the actions of their citizens in space, he argues that it will be difficult, if not impossible, for a company to operate in space without supervision and accordingly the business of space will be subservient to national interests and will face major regulatory controls.

Hertzfeld characterizes US government policy toward commercial space as being confused in terms of the signals it provides to the space industry, to foreign governments and to potential competitors. He explains this confusion by pointing to the important role of space in national security and a goal of reserving some space capabilities, whether commercially- or government-owned, for national purposes; a rapidly changing industry that has not as yet reached commercial maturity; the use of space assets for international political purposes; and changes in government policy over time concerning competition and de-regulation. He notes that most other nations have developed space capabilities and space programs to encourage and subsidize economic growth through cutting-edge technological developments as well as to create jobs. The level of US space manufacturing industry employment from 1997 to 2007 is shown in Figure 3.2.

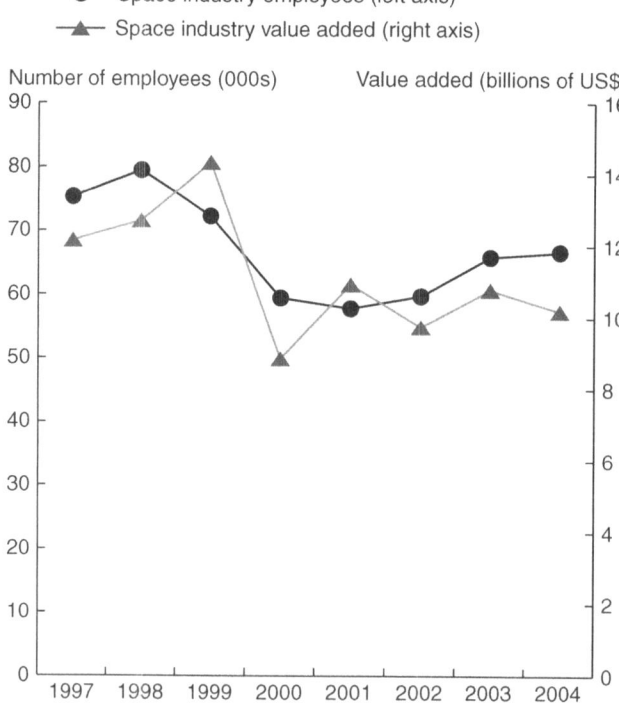

Source: OECD (2007).

Figure 3.2 US space manufacturing industry employment and value added, 1997–2004

Since 1960 there have been eight major Presidential Documents on space policy, the most recent from the Obama Administration. These policies reflect changing technological, political and economic conditions (Mustafa 2004). Hertzfeld (2007) finds that it is clear from a very rudimentary count of words in these documents that the economic and commercial aspects of space became important policy considerations only in the 1980s (see Figure 3.3).

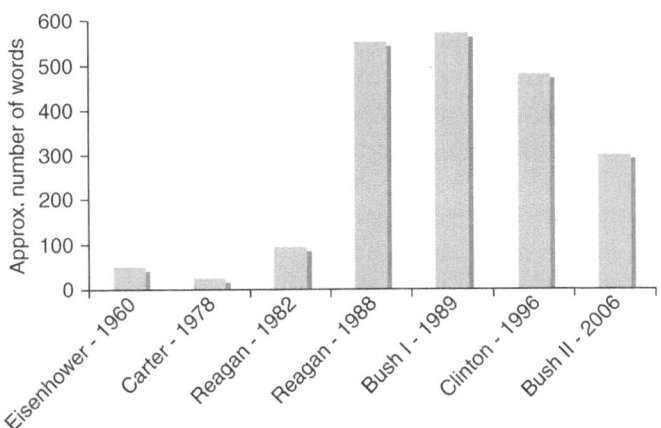

Source: Hertzfeld (2007).

Figure 3.3   Commercial space in presidential space policy

Space policy originated during the Cold War, with an emphasis on security and emerging ahead of the Soviet Union in the space race. Hertzfeld (2007) finds that there was a general recognition in the Eisenhower Policy that the design and development of space equipment would stimulate the economy via job creation and the possibility of spin-off products entering the economy, and that it also recognized the future potential economic aspects of two civilian applications of space technologies: communications and meteorology.

During the 1980s, commercial companies that had been solely government contractors for space equipment were branching into independent offerings of space components and systems (Hertzfeld 2007). Meanwhile, in the USA the era of deregulation had begun, with the ascendancy of Ronald Reagan to the Presidency. The Reagan Administration mandated that the government should 'obtain economic and scientific benefits through the exploitation of space, and expand United States private-sector investment and involvement in the civil space and space-related activities', and the succeeding Bush Administration followed in a similar vein.

The Clinton Administration further promoted commercial space. It introduced policies to promote remote sensing activities and to maintain American competitiveness in the remote sensing sector. The Clinton era also saw one of the first major initiatives in space involving a public/private partnership, in the development and operation of a new, reusable space transportation system. A policy directive issued in 1996 recognized clearly that the private-sector investment in US GPS technologies and services was important for economic competitiveness, and the policy encouraged continued private activity in this area, subject to issues of national security (Hertzfeld 2007).

While the most recent Bush Administration continued to promote and encourage commercial activity, the overall policy document issued in August 2006 had less emphasis on commercial considerations and an increase in emphasis on national security considerations. None the less, commercial and entrepreneurial engagement in the business of space continued.

The current Obama Administration believes the United States should maintain its international leadership in space while at the same time inspiring a new generation of Americans to dream beyond the horizon:

> When I was growing up, NASA united Americans to a common purpose and inspired the world with accomplishments we are still proud of. Today, NASA is an organization that impacts many facets of American life. I believe NASA needs an inspirational vision for the 21st Century. My vision will build on the great goals set forth in recent years, to maintain a robust program of human space exploration and ensure the fulfillment of NASA's mission. Together, we can ensure that NASA again reflects all that is best about our country and continue our nation's preeminence in space. (Obama, quoted in Cantrell 2008)

As president, Obama has established a robust and balanced civilian space program. Within this view, NASA not only should inspire the world with both human and robotic space exploration, but it should also again lead in confronting the challenges human beings face on Earth, including global climate change, energy independence and aeronautics research. In achieving this vision, Obama is attempting to reach out to include international partners and to engage the private sector to amplify NASA's reach. Obama believes that a revitalized NASA can help America to maintain its innovation edge and contribute to American economic growth. There is currently no organizational authority in the Federal government with a sufficiently broad mandate to oversee a comprehensive and integrated strategy and policy dealing with all aspects of the government's space-related programs, including those being managed by NASA, the Department of Defense, the National Reconnaissance Office, the Commerce Department, the Transportation Department and other federal agencies. This was not always the case. Between 1958 and 1973, the National Aeronautics and Space Council oversaw the entire space arena for four presidents; the Council was briefly revived from 1989 to 1992. Obama is planning to re-establish this Council reporting to the president. According to his vision, the Council will oversee and coordinate civilian, military, commercial and national security space activities. It will solicit public participation, engage the international community, and work toward a twenty-first-century vision of space that constantly pushes the envelope on new technologies as it pursues a balanced national portfolio that expands our reach into space and improves life here on Earth.

Hertzfeld (2007) concludes that, overall, space policy directives have slowly been transformed from a Cold War emphasis that marginalized the economic and commercial implications of space activities into a truly integrated policy that recognizes the maturity of many space applications, sophisticated industrial capabilities, the globalization of space technologies, and the importance of the space infrastructure for civilian uses and security concerns. More recently, following the events of 11 September 2001 (9/11) and their aftermath, US space policy continues to promote the business of space but security considerations have also gained prominence.

Unlike in the Cold War era, when the Soviet Union was the only rival to US space power, today competition is emerging from many parts of the world. The remaining sections of this chapter consider the other major players – Russia and the ESA, and the emerging players – China, India, Japan, Brazil, Korea, Israel, Pakistan, Saudi Arabia and Iran.

Russia

Apart from the USA, the Soviet Union was the only country in the world which pursued the entire spectrum of space research and applications – scientific, commercial and military. From the late 1960s onward, the number of Soviet space launches by far surpassed the combined launch rate of all other countries in the world (see Figure 3.4).

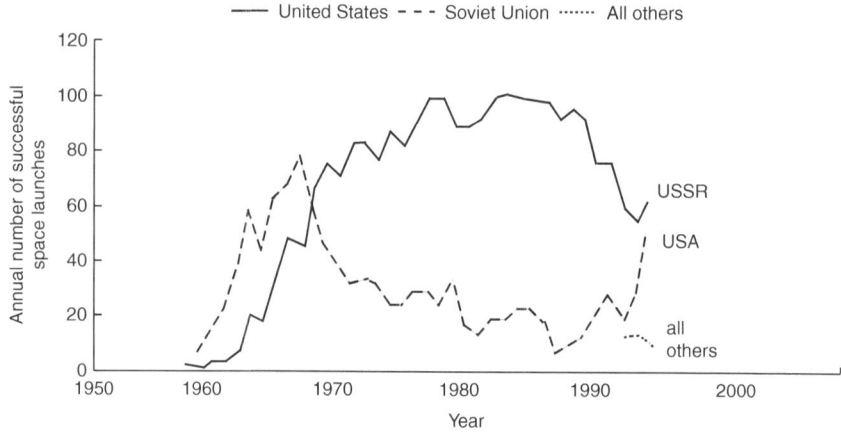

Note: Launch failures are not included because of incomplete data on Soviet launches. Number of launches in 1993 is estimated.

Source: Tarasenko (1996a).

Figure 3.4 Annual number of successful space launches worldwide

The Soviet space program began with intercontinental ballistic missile (ICBM) development in the 1950s. The rocket and space programs were considered to be an integral part of the Cold War rivalry with the United States. As such a powerful rocket and space industry, versatile R&D facilities, and an extensive infrastructure to support both missile testing and space operations were put in place (see Figure 3.5).

Source: Tarasenko (1996b).

Figure 3.5 Rocket test ranges and major rocket production facilities, Russia, Ukraine and Kazakhstan

Hundreds of enterprises with a cumulative workforce of more than one million people were estimated to have participated directly in the Soviet missile and space programs (Tarasenko 1996b). An important difference in the Soviet space program compared to that of the USA was that there was no clear separation of military and civilian space activities. However, the upheaval that accompanied the end of the USSR extended into its space pursuits. Ultimately the space program, including the various assets and enterprises of the former Soviet Union, ended up largely under the control of Russia. Over the period from the collapse of the Soviet Union, there has been an increasing emphasis on the exploitation of Russia's extensive space capabilities, including some hitherto dedicated military systems, for various civil applications. Despite the disruption caused by the break-up of the Soviet Union and the subsequent economic crisis, the decline in space activity as measured by the number of space launches was halted with the transfer of the space program to Russia, with the number of space launches returning to the level of the late 1980s.

The commitment of Russia to space can also be measured by the diversity of satellite constellations that remain functional, with Russia keeping more than 20 operational satellite systems (Tarasenko 1996b). At the end of 1992, the total constellation of operational Russian spacecraft consisted of about 140 satellites, more than the year before. By 2010, Russia expected to maintain 160 to 180 spacecraft, representing about 30 satellite systems, for scientific, economic and military purposes (Tarasenko 1996b).

Tarasenko has classified Russian space systems according to the missions performed. (see Figure 3.6). These systems can be sub-divided into space weapons; space surveillance and intelligence systems; support systems; and scientific systems.

Focusing on navigation, two generations of satellite navigation systems are currently in use. First-generation systems provide naval and trade vessels with the opportunity to determine their positions within one or two hours anywhere on the globe. The second-generation system (known as GLONASS in the West, but named Uragan by the Russian military) is similar to the US Navstar Global Positioning System. Operational deployment started in 1989 and has now resulted in a constellation of 15 satellites, making it a complete system.

Many military space systems have civil applications. For example, the same space launch services are used for civilian and military payloads. Communications, navigation and weather data are necessary for both civilian and military operations.

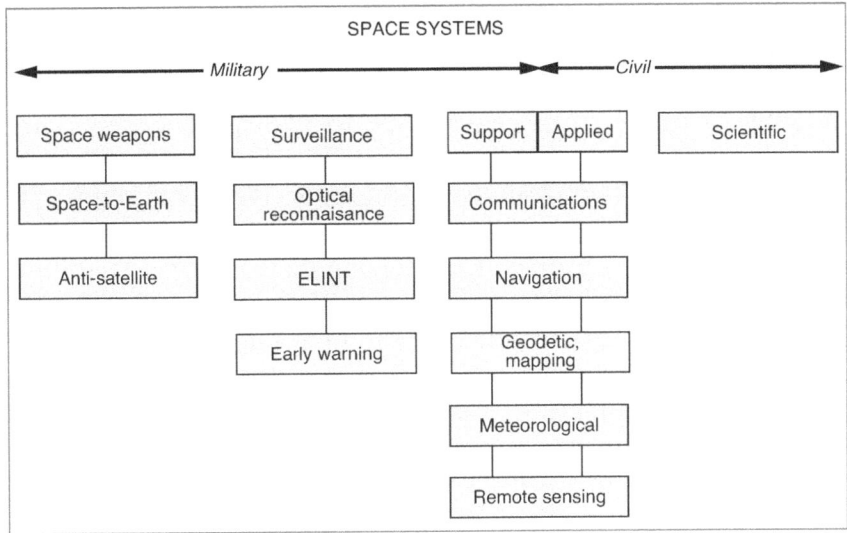

Source: Tarasenko (1996b).

Figure 3.6 Classification of space systems by missions performed

Soviet meteorological satellites and some navigation satellites have already been utilized by both military and civilian users. In recent years, in the face of resource constraints, there has been some relaxation of security restraints, with the result that many proposals have been made to Western businesses by the Russian space industry. Among the assets made available for commercial use were Russia's extensive launch support infrastructure, the former Soviet system of strategic communications and Earth observation satellites. There have been some examples of joint ventures between Russian and Western enterprises.

Europe

The importance of space to Europe is defined within the Lisbon Strategy, which aims to make the Union the most advanced knowledge-based society in the world. In 2007, the European Union further extended its scope and diversity by enlarging to include 27 member states and 500 million citizens. Naturally, this means adapting existing instruments and technologies, or creating new ones, to manage the EU successfully.

Space technologies are particularly useful in this respect because of their capacity to collect and distribute information at any location for every citizen. The ways that space affects the lives of citizens range from health care to transport systems, and most Europeans are not even aware of the fact that many of such provisions stem from technologies derived from space. The need to reinforce Europe's capabilities in space is becoming ever more apparent as Europeans are relying increasingly on satellites for communication, navigation, monitoring the environment, developing innovative technology and increasing scientific knowledge, as well as numerous other space-related technologies. Daily personal benefits include weather forecasting and satellite TV. In Europe, 1,250 television programs are broadcasted by satellite to 100 million homes. Given the importance of space activities to the standard of living of Europeans, space developments were consequently included in the EU Constitutional Treaty.

Investigating the benefits that have emerged from leveraging space activities provide clear incentives for the ESA to strive to be a competitive force in this sector. Numerous benefits that improve daily life, such as environmental security, better telecommunications, the creation of new jobs, and navigational assistance stem from space-related activities and thus make space involvement an essential area of investment for European countries (Eurospace 1995).

Until the 1960s, space exploration was originally driven by nations' individual motives and endeavours. Over time, it was recognized that a better collaborative effort was needed among all the European countries that would benefit. In 1964, the European Space Research Organisation (ESRO) and the European Launcher Development Organisation (ELDO) were created in an attempt to develop a collaborative framework. In 1975, these two organizations combined to form what we now know today as the European Space Agency (ESA). Since then, the ESA has been responsible for coordinating the collective efforts of the European member states in the space arena.

The European Commission (EC) and ESA share a common aim: to strengthen Europe and benefit its citizens. ESA is an intergovernmental organization with no formal organic link to the EC. However, in recent years the ties between the two institutions have been reinforced by the increasing role that space plays in strengthening Europe's political and economic position, and in supporting European policies. To facilitate relations between the two organizations, the ESA set up a liaison office in Brussels, the location of the headquarters of the EC.

The overall structure of the European space industry comprises three layers: the EU; intergovernmental organizations such as the ESA; and the national space agencies. The total business of the European space industry has a double nature: institutional and commercial (Eurospace 1995). Space activities are gradually evolving from those that are publicly financed to commercial ventures. The emergence of the ESA allowed companies to foresee income streams long-term as a result of industrial policy. The ESA also removes much of the risk associated with the space industry by returning most of each nation's budgetary contribution in the form of industrial contracts, known as geographical return.

The commercialization process in Europe has also been driven by increased technical maturity in space technologies as well as the overall globalization of the space market. The open commercial market represents around 30 percent of global space activities amounting to a turnover of €3 billion. The remainder constitutes institutional demand. Attention to the institutional sector has been low in the EU, providing the USA with a very strong competitive advantage. To maintain long-term sustainability and fully reap the benefits of the space sector, Europe needs to create a more robust institutional market. The USA bases its market heavily on national security and defense, which is closed to foreign suppliers (Eurospace 1995).

One of the most important strategic objectives of any economy similar to the EU is economic prosperity. To secure growth and job creation, the Union must successfully facilitate the transformation of its society into a knowledge-based one, and space adds value to a knowledge-based society. Research provides access to advanced technologies and services, making Europe more competitive. European space activities employ around 40,000 people. Indirect employment is estimated at 250,000. These numbers are increasing as space technologies develop.

The ESA aims to nurture small and medium-sized enterprises (SMEs) in their early development. As they generally grow and develop after the initial assistance of capital, they thereby relieve ESA of some responsibility, which can then be transferred to the creation of other start-up companies. However, prosperity and growth cannot develop without investment. The ESA budget is the next important consideration. The annual ESA budget is currently €3 billion, of which €400 million is allocated to technology and R&D. Total operating expenditure increased by 5 percent in 2006 as a result of the reinforced emphasis on technological research activities in preparation for future missions. The ESA's mandatory activities represented 28 percent of the total expenditure, while their optional programs spent 69 percent. Member states declare a voluntary subscription on an annual basis to the budgets for mandatory and optional activities. The ESA currently spends the majority of its budget on Earth observation, human space flights and navigation. The operating income for the ESA for 2006 was €3449 million, of which 80 percent was made up of contributions from its members.

China

On 8 October 1956, the Central Committee of the Communist Party of China, presided over by Mao Zedong, established the Fifth Research Academy of the Ministry of National Defense to develop the country's space effort. This was the official beginning of the People's Republic of China's (PRC) space program. Four years later, on 5 November 1960, China launched its first rocket, becoming the fourth country after Germany, the USA and the USSR to enter space.

Since then, China has aimed to become a recognized international space power, thereby advancing its military capabilities. With the successful completion of the Shenzhou 4 flight on 15 October 2003, China's ambitious space program developed dramatically and it quickly accomplished its first manned orbit of the Earth. This flight (Shenzhou 5), a milestone in China's space development, should not be considered an end in itself, but rather the entry permit to the space-power club of the USA and Russia (Liao 2005).

According to Liao (2005) China's space program has gone through five distinct periods. In the first period, 1956–66, the Chinese established a space program despite the trauma caused by Mao Zedong's 'Great Leap Forward'. Mao sought to push China to follow the USSR and the USA in seeking missile and space-launch capabilities. At that time, China had few resources, either technological or economic, and was a far poorer nation than either the USA or the USSR (Fairclough 2007).

In the second period, 1966–76, it began to devote significant resources to the development of satellites following the launch of its first rocket in 1960. The country's space program maintained a progressive course, even though sectors of Chinese society were being torn apart by the Cultural Revolution. China launched its first satellite, Long March 1 (Dong Fang Hong 1 (DFH 1)), on 24 April 1970, making it the sixth nation to launch its own satellite into orbit (after the USSR (Sputnik 1, 1957), the USA (Explorer 1, 1958), Canada (Alouette 1, 1962), France (Astérix, 1965), and Japan (Ohsumi, 1966). Since DFH 1 in 1970, China has launched some 60 satellites in five primary categories, which support both civil and military efforts. These are, namely, retrievable remote sensing satellites, Fanhui Shin Weixing (FSW); communication satellites, Dong Fang Hong (DFH); meteorological satellites, Feng Yun (FY); scientific and technological experimental satellites, Shi Jian (SJ); and Earth-imaging satellites, Ziyuan. During the second period, China developed an indigenous family of liquid-fueled space-launch vehicles that are competitive with Western launchers, a large national space research effort, and an extensive satellite industry. In addition, it made space remote sensing a priority and has developed its own communications and navigational satellites. And, despite China's satellite industry lagging behind that of the USA and Europe, joint ventures with foreign firms over the decade 1966–76 helped it to improve its satellite manufacturing capabilities (Liao 2005).

The third period, 1976–86, was an ambivalent period for the space program, as China's recovery from the Cultural Revolution proceeded slowly under Deng Xiaoping's leadership. Mao had focused the program on national prestige and national security, but Deng's Four Modernizations program placed the highest priority on economic and scientific efforts to help develop the economy. Therefore China's space budget was trimmed to meet more modest ambitions. With the reductions in space spending in the early 1980s came Beijing's authorization for China's space agencies to generate income from external sources. On 29 January 1984, the Long March 3 launch vehicle inaugurated the Xichang facility in south-central China. Following this successful launch, China began to offer Long March launchers to the West. In effect, China's space program was shifting its orientation from the defense sector to the civilian/commercial sector. For example, the National High Technology Research and Development Project of China (Project 863) was launched in March 1986 by the central government with the aim of enhancing China's international competitiveness and improving its overall R&D capability in high technology. Project 863 focused on the leading strategic and upcoming high technologies that would benefit China's long- and medium-term development and security. In addition, the creation of Project 863 was an effort to place China in a position to concentrate its space program on practical applied satellites. Since then, high technology development has served not only military and political purposes, but also civilian and commercial uses.

At the start of the fourth period, 1986–96, China was getting little response to the launch services it offered until a series of events occurred in 1986, starting with the disastrous loss of the US Space Shuttle Challenger. Then two of the USA's other leading rocket launchers, a Titan and a Delta, spectacularly exploded, and Europe's Ariane went down. The China Great Wall Industry Corporation (CGWIC) had been actively marketing Chinese launch services in the hope that the flexible Long March family of launch vehicles would prove attractive to the international market (Liao 2005).

The first launches for paying customers involved experimental payloads using the Long March 2 – the first for a French company (Matra) in 1987 and then for a German consortium (Intospace) in 1988. China's Space Leading Group (SLG) was established in 1991 and has overseen and coordinated all space activities in a broad policy-making role. The Chinese National Space Administration (CNSA), established in 1993, is the executive agency for space functions, responsible to the State Council. In this period, aerospace technology was able to contribute significantly to other national economic development efforts and was highlighted in the Five-Year Plan (1991–5).

The fifth period began in 1996 with the Five-Year Plan (1996–2000). Since then, China's commercial launches have been shared between two government organizations – the Commission of Science, Technology and Industry for National Defense (COSTIND); and the China Aerospace Corporation (CASC). During this period China clearly benefited from US, Russian and European openness in space technology (Liao 2005). For example, China and the EU have recently agreed on Chinese participation in the Galileo navigation satellite program (alongside Canada and Israel). This collaboration will allow China to develop a more sophisticated understanding of navigational satellites. Galileo is an example of how China has used foreign partnerships to speed up its indigenous space effort – not through pirating technology, but by participating in and learning from the experience of other programs.

It is clear that a great deal of the technology now being applied to commercial space systems can be carried over to military systems. In this regard, the USA is concerned that China will employ dual-use and pirated or transformed technologies in both commercial and military space programs. This period culminated in the historic Shenzou 4 launch. The first Chinese manned space flight program took place with Yang Liwei's successful 2003 flight aboard Shenzhou 5. This success made China the third country to independently send humans into space. But what is the next space development for China after manned space flight? The first feasibility study for lunar exploration was completed in 1995, and the development of lunar exploration technology was spelled out in China's first White Paper on space. The China State Council's ruling cabinet leadership approved the eleventh Five-Year Plan (2006–10) for space on 10 May 2007 (Vick 2009).

The Chinese government had previously released a White Paper report on the results of the previous Five-Year Plan

(2001–5) and the immediate future plans for the Five-Year Plan for 2006–10. Within that report emphasis was placed on satellite remote sensing expansion, direct broadcast satellites, meteorological satellites and navigation satellite developments. It also emphasized the development of new sea monitoring satellite series and a small constellation satellite series. The space plans were subsequently approved by the Chinese Communist Party central committee of about 350 members before the Party Congress approved it. China's space program aims to emphasize three areas of development: navigation, remote sensing and communication satellites; all having dual purpose require-ments for the civil and military sectors of the national space effort. One of the efforts is concentrated around the BeiDou GPS class satellite competitor to the GPS, GLONASS and planned Galileo satellite systems.

In 2009, China planned to launch as many as 15–16 satellites spread out over a wide-ranging series of manned and unmanned programs. China also announced the development of a new satellite production infrastructure in the city of Shenzhen called the Shenzhen Aerospace Spacesat Co. Ltd. The state plan is for the organization to develop six to eight satellite designs with the goal of producing five to six of then each year. The high technology facility is expected to cost 160 million yuan (Vick 2009).

Korea

Korea began to participate in the space sector only in the early 1990s. Its first project was a micro-satellite named KITSAT 1, while its first sounding rocket project, KSR-1, with a first-stage solid propellant, had begun in 1989. With these two small scientific projects, Korea stepped into the sophisticated and technology-intensive space industry.

The space industry is a symbolic industry through which Korea will be able to prove its national technological capability as well as its national power (Hwang 2006). Korea is a latecomer to this particular industry, deciding to participate almost 50 years after the launch of the first artificial satellite. The space industry has very different characteristics from the other conventional mass production industries with which Korea has successfully caught up. Technology accumulation through learning by doing is very lengthy, as the production unit remains at almost one per project. However, since the year 2000 Korea has improved its technological capabilities in satellite development, as well as making progress in space launch vehicle technology.

According to the Space Development Promotion Act enacted in May 2005, the supreme government body for deciding space policy in Korea is the National Space Committee, which is placed under the control of the President and chaired by the Minister of Science and Technology. The committee consists of around 15 members, including nine ministers of related ministries. The Ministry of Science and Technology (MOST) is the major government body for formulating and executing the national space development plan. Among the related ministries, the Ministry of Commerce, Industry and Trade (MOCIE) is mainly concerned with fostering manufacturing industries; the Ministry of Information and Communication (MIC) deals with the information and communications sector, including satellite broadcasting and communication; the Ministry of Construction and Transportation (MOCT) is responsible for CNS/ATM (Communications, Navigation and Surveillance Systems for Air Traffic Management) and land development; and the Ministry of Maritime Affairs and Fisheries (MOMAF) is involved in monitoring the ocean environment and marine ecosystem, production of fisheries information and so on.

The main space development research institute in Korea is the Korea Aerospace Research Institute (KARI), established in 1989 under the supervision of MOST. Space activities in Korea have mainly been carried out by government research institutes in cooperation with foreign companies (Hwang 2006). Furthermore, the number of development projects connected to production units have been very limited thus far, which is why Korean industries have been relatively less developed. Recently Korean satellite development projects have been growing fast in terms of numbers of projects as well as the monetary base and this will provide more opportunities for Korean companies to be involved in space development projects. Korean companies have increased their technological capabilities through learning by doing. Other Korean companies have been working as subcontractors for the development of the Korean Sounding Rocket and Space Launch Vehicle Program.

Satrec Ltd is a technology-oriented venture company which developed Lazak-Sat, a 100 kg class micro-satellite for remote sensing, for Malaysia in 2005. It is the only Korean company to date to have developed and exported a satellite without outside technological assistance.

Korea established its first National Space Development Mid- and Long-Term Basic Plan in 1996 and revised it in 2000 and 2005. The plan incorporates basic activities in space up to 2015. The long-term objectives of space development are to acquire the independent technological capabilities for space development and to join the top 10 countries in the space industry by competing in the global market (Hwang 2006).

The mid-term objectives were more specific. Korea accomplished the first target by acquiring the capability to launch micro-satellites by 2007. Second, by the year 2010, a low Earth orbit (LEO) multipurpose satellite was to be developed independently. Finally, the technical basis to compete in the global space market would be established. In order to accomplish these objectives, Korea planned to develop 13 satellites (four satellites in the initial phase) by the year 2010 (seven multipurpose satellites, four scientific satellites and two geostationary satellites). When the mid-term plan is completed, Korea will have acquired the capability to develop LEO multipurpose satellites domestically. Multi-purpose satellites will meet the national need for monitoring the ground, the ocean and the environment as well as satisfying the public needs for continuous satellite data. Science and technology satellites will perform preliminary research in core technologies necessary for the development of the multi-purpose satellites, as well as performing space science experiments. The Communication, Ocean and Meteorological Satellite (COMS) will be used to acquire the technology to develop a geostationary satellite locally and satisfy the need for satellite communications, ocean monitoring and meteorology services.

Brazil

The Brazilian Space Agency (abbreviated in Brazilian Portuguese as AEB) was established in 1994 as a civilian authority within the direct purview of the Executive Office of the President of Brazil. It is responsible for pushing forward Brazil's space activities and for coordinating the national and international cooperation necessary to help further the country's strategic goals in space.

The National Policy on the Development of Space Activities (PNDAE) establishes the major principles, objectives and guidelines for Brazilian space activities. The policy outlines several specific objectives: namely, to establish within Brazil scientific and technical competence in the space arena that will make it possible for the country to act with real autonomy in some well-identified situations, including the selection of technological solutions to Brazilian problems, and pursuing its national obligations under relevant international negotiations, agreements and treaties; to promote the development of space systems, and related ground infrastructure, that will provide data and services desired by the country; and to prepare the Brazilian industry to participate and become competitive in the global market for space-related goods, services and applications (Silva 2005).

The National Institute for Space Research (INPE) was founded in 1961 to enable Brazil to benefit from new developments in space science and technology, and in particular to increase Brazil's autonomy in strategic areas, by providing a means for industry to become competitive in the space sector and by encouraging the development of a national space technology capability. INPE's main goals are the fostering of scientific research, technological applications, and to qualify personnel in the fields of space and atmospheric sciences, space engineering and technology.

The budget for Brazilian space activities has been steadily increasing, albeit from a low base, as part of a long-term governmental policy to increase the country's investment returns on science and technology (Zhao 2005).

Commercial activities include Embratel's Brasilsat tele-communications network, developed by Spar (of Canada) and launched on Ariane in the mid-1980s. A second generation of satellites was launched in the early to mid-1990s. A constellation of twelve EO satellites (ECO-8) is currently under development, with an estimated cost of some US$600 million. There are also plans to open the Alcantara launch center to foreign users. In 1993, the Brazilian Aerospace Industries Association was established to represent the interests of Brazilian companies working in aerospace (and space) engineering. While Empresa Brasileira de Aeronbutica – usually called Embraer – dominates the Brazilian aerospace/space industry, other companies have established expertise in key areas, including rockets and missiles, avionics and other electronics and composite materials (Silva 2005).

Compared with other countries, Brazil is a modest player on the global space scene. As available resources have not allowed the nation to advance in this sector, partnerships open a window of opportunity for Brazil to take part in a much wider range of projects than would be possible if these had to be funded by the country alone. International cooperation is considered to be the best way to participate in strategic projects at a lower cost, and to have access to new technologies developed abroad. To keep these objectives on track, the Brazilian government established a national space policy with the following priorities: 'emphasis on applied sciences and applications, in particular microgravity research; participation in the ISS and in projects of space infrastructure, aiming at carrying out scientific and technological experiments' (Zhao 2005).

The PNDAE covered a 10-year period, from 1998 to 2007. The program consisted of eight major initiatives: Space Applications; Satellites and Payloads; Satellite Launching Vehicles and Sounding Rockets; Space Infrastructure; Space Sciences; R&D on Space Technologies; Training and Development of Human Resources; and Support to the Qualification of the National Space Industry.

Notable activities of the Brazilian National Space Program include the three small data collection satellites developed by Brazilian engineers: SCD-1, was launched in 1993 and remains operational; SCD-2 was lost on the first vertical launching system (VLS) launch; and SCD-3 was launched in 2000. In addition, since 1988, China and Brazil have been collaborating on a program to develop two remote sensing satellites. The first, CBERS (China–Brazil Earth Resources Satellite) satellite, CBERS 1 was launched in 1999 and was the first in a program of remote sensing satellites designed for global coverage using optical visible and infrared cameras. Initially a two-satellite system, China and Brazil have now agreed to develop two second-generation satellites, CBERS 3 and 4. The two countries have also agreed to explore the feasibility of the joint development of a geostationary meteorological satellite and a telecommunications satellite, based on the CBERS satellite bus (Zhao 2005).

In addition, SACI, a micro-satellite launched in 1999 carrying four experiments conceived by Brazilian scientists and their foreign partners, is providing much useful information. Other small scientific satellites are expected to be launched in the short and medium term. Finally, two small Earth-observation satellites, SSR 1 and SSR 2, projected to operate in an equatorial circular orbit at an altitude of 900 km, were developed. The first was launched in 2000, and the second launched in 2003. A low Earth-orbit equatorial satellite constellation, aimed at providing low-cost communications to remote areas around the Equator was under consideration but at the time of writing it has not yet taken place.

As a result of the cooperation agreement signed between NASA and the AEB, the latter will be responsible for the development and provision to NASA of equipment for the International Space Station (ISS) program. In exchange, the AEB will receive rights from NASA's allocation to use the ISS (Zhao 2005).

Since the early 1970s, Brazil has been engaged in a long-term launcher development program which began with the development of SONDA, a successful family of sounding rockets. The Alcantara Launch Centre (CLA) is capable of launching solid-fuel sounding rockets and research vehicles, as well as satellites into low Earth orbit. Located on Brazil's north-eastern coast, near the Equator, CLA's geographical position increases its safety and enables lower launching costs. In the coming years the CLA is expected to enlarge its capacity and become commercially competitive, for both national and international users, and Infraero, the organization responsible for managing Brazil's airports, has been appointed to administer those areas of CLA that will be open to foreign launchers and operators.

India

Despite being a developing economy with its attendant problems, India has developed effective space technology and applied it successfully for rapid development so that today the country is offering a variety of space services globally (Morring and Neelam 2004).

During the formative decade of the 1960s, space research was conducted by India mainly with the help of sounding rockets. The Indian Space Research Organisation (ISRO) was formed in 1969. Space research activities were given a further impetus with the formation of the Space Commission and the Department of Space by the Government of India in 1972. The ISRO was brought under the Department of Space in the same year. In the history of the Indian space program, the 1970s were the era of experimentation, during which satellite programs were conducted. The success of these programs led to an era of operationalization in the 1980s, during which operational satellite programs such as INSAT and IRS came into being. At the time of writing, INSAT and IRS are the major programs of the ISRO (ISRO 2009).

To launch its spacecraft independently, India has a robust launch vehicle program, which has matured to the point of offering launch services internationally. The Antrix Corporation, the commercial arm of the Department of Space, is marketing India's space services globally. Fruitful cooperation with other space-faring nations, international bodies and the developing world is one of the main characteristics of India's space program (Morring and Neelam 2004).

The most significant milestone of the Indian Space Program was during the year 2005/6 with the successful launch on

5 May 2005 of the ninth flight of the Polar Satellite Launch Vehicle (PSLV-C6) which placed two satellites – the 1,560 kg CARTOSTAR-1 and the 42 kg HAMSAT – into a predetermined polar sun-synchronous orbit (SSO). Coming after seven launch successes in a row, the success of PSLV-C6 further demonstrated the reliability of the PSLV and its ability to place payloads weighing up to 1,600 kg into a 600 km high polar SSO.

The successful launch on 22 December 2005 of Insat 4A, the heaviest and most powerful satellite built by India to date was the other major event of the year 2005/6. Insat 4A is capable of providing direct-to-home (DTH) television broadcasting services (ISRO 2009).

In addition, the setting up of the second cluster of nine Village Resource Centres (VRCs) was an important ongoing initiative of the Department of Space during the year 2005/6. The VRC concept integrates the capabilities of communications and earth observation satellites to provide a variety of information emanating from space systems, and other IT tools, to address the changing and critical needs of rural communities. More than 400 VRCs have been set up, to provide rural communities with information on natural resources, land and water resources management, and telemedicine. Using Insat, around 400 hospitals in remote and difficult-to-reach locations have been integrated into the telemedicine network. The ISRO is also planning an exclusive satellite for boosting rural connectivity. India convincingly demonstrated its capability for a deep space mission with the smooth insertion in November 2008 of its maiden lunar probe Chandrayaan-1, launched in October 2008, into a 100-km orbit around the moon (ISRO 2009).

For a developing nation that began its space journey with the test firing of a 9 kg sounding rocket from the fishing hamlet of Thumba near Thiruvananthapuram in the state of Kerala in November 1963, Chandrayaan-1 was a success on a shoestring budget. With a cost of less then Rp4 billion (US$83 million), Chandrayaan-1 is considered to be the most inexpensive lunar probe ever launched – its cost is nearly one-third of China's Chang'e 1 and one-sixth of Japan's Kaguya. 'With a minuscule budget, we have mastered cutting-edge technology in space,' says ISRO chairman Madhavan Nair (Radhakrishna 2009).

Nair says Chandrayaan-1 is part of India's long-term vision. He believes Chandrayaan-1, with its 'unique combination of payloads', will facilitate comprehensive mapping of the lunar surface for the first time. Key objectives of Chandrayaan-1 include identifying and mapping mineral resources, looking for signs of ice water and confirming the presence of helium 3, a clean and green energy source. 'The Indian Moon mission should be seen beyond the scientific results it provides. Studies have shown that the Moon could serve as a source of economic benefit to mankind and be of strategic importance,' says Mylswamy Annadurai, Chandrayaan-1 project director. The ISRO says that Chandrayaan-1 is expected to complete the mapping of the moon by the end of 2010 (ISRO 2009).

The ISRO has initiated work on a Rp4.2 billion Chandrayaan-II mission, set for launch in 2011–12 atop the three-stage Geosynchronous Satellite Launch Vehicle (GSLV). The Indo-Russian Chandrayaan-II will feature a lander with a rover, which will be used to collect samples of lunar rocks and soil, subject them to chemical analysis and then transmit the data to the main orbiter. Beyond Chandrayaan-II, India is looking at a 'sample return mission to the Moon'. Nair has said, 'if we find mineral resources on the Moon, the next logical step will be to collect and bring them back to Earth'. As this would need a massive lift-off rocket, the ISRO started to develop a semi-cryogenic launch vehicle working on refined paraffin and that was expected to be ready in six years. The ISRO is also aiming for recoverable and reusable launch vehicles as part of its long-term strategy of making access to space affordable and routine (Radhakrishna 2009).

Also on the ISRO's agenda are plans to launch a Mars orbiter, to send a probe to Venus and to explore the asteroid belt. There are plans to land a spacecraft on an asteroid belt and to send a probe to fly past a comet before 2020. The Rp120 billion Indian manned mission is now a priority area for the ISRO. An Indian spaceship with two or three crew members is planned to be launched into a 400km near-Earth orbit by a GSLV MkIII vehicle in 2015 (ISRO 2009).

For the Bangalore-based Antrix Corporation, the successful launch of Chandrayaan-1 using an augmented version of the Polar Satellite Launch Vehicle promises more customers for its PSLV 'cost-effective launch service'. The PSLV has to date launched 30 satellites, including 16 from overseas. In April 2008, the PSLV set a record by launching 10 satellites at the same time. Eight were microsatellites, weighing from 3–16 kg, from Canada, Europe and Japan. Recent contracts won by Antrix include launching Algeria's Alsat-2A and Italy's IMSAT spacecraft on board a PSLV in 2009. The company also has in hand contracts to launch a microsatellite from Singapore's Nanyang Technological University, and CubeSat, a three-satellite package from the Netherlands, on the PSLV. The launch of GSLV with an Indian upper cryogenic stage and GSLV MkIII promises further commercial opportunities for Antrix. The ISRO launched a GSLV flight with a homegrown cryogenic engine – as a replacement for the Russian-supplied stage – in 2010.

Antrix, which has been supplying components and sub-systems to global satellite builders, has also delivered the W2M satellite to Eutelsat, while sales of satellite images is a growth area for the company. Revenue from satellite data sales accounted for 10 percent of its turnover of Rp9.4 billion in 2007/8, with its market expanding beyond Europe and the USA to include Australia and Russia. Asian and African countries now source remote sensing data from Antrix.

Since its inception, the focus of India's space program has been on exploiting space technology to accelerate the pace of national development. With little outside assistance, the ISRO has built an impressive base. The remote-sensing Earth observation constellation is made up of seven satellites, and the Insat communications constellation comprises 11 spacecraft supporting activities including agriculture and resources exploration, fisheries, disaster management, weather forecasting and TV broadcasting: 'Of our budget of less than $1 billion, 80% is being used for societal benefits,' says Nair (Radhakrishna 2009).

The success rate of water exploration schemes in India has risen by 50–80 percent following the use of satellite data. Similarly, a system for weather prediction and disaster warning is in place.

In the 1960s, Vikram Sarabhai, the architect of the Indian space program, observed that (Radhakrishna 2009):

> we don't have the fantasy of competing with economically advanced nations in the exploration of the Moon or planets or manned flights. But we are convinced that to play a meaningful role nationally and in the community of nations, we must be second to none in the application of advanced technologies to the real problems of man and society which we find in our country.

India's scientific community perceives this shift as a justifiable development in keeping with the nation's emergence as an economic power and a technological hub. ISRO chairman Nair is quick to point out that 'as far as space is concerned, India is considered a developed country' (Radhakrishna 2009).

Iran

Iranian efforts to exploit space began under the last shah, Mohammad Reza Shah Pahlavi who tried to improve his country's scientific standing. In 1959, Tehran became a founding member of the United Nations' Committee on the Peaceful Uses of Outer Space (UNCOPUOS). Iran has considered becoming a space power as a vehicle for modernity. Some of the goals it listed at a 2002 UNCOPUOS meeting reinforces this perception:

1. Commercializing space programs for Earth observation, and predicting environmental changes.
2. Promoting international cooperation based on concepts of joint benefits.
3. Encouraging space efforts in the private sector to increase awareness within the public of space and incorporate related initiatives into daily lives.
4. Developing a mastery of space science and technology directed to assist in the development of space programs and commercial projects.
5. Increasing interest in space programs among the youth, who will play a notable role in the country's future.

Iran sought to accomplish these and other broad objectives in order to become more technologically advanced. On 5 January 2003, Rear Admiral Ali Shamkhani, the country's former defense minister, stated that, within eighteen months, 'Iran will be the first Islamic country to penetrate the stratosphere with its own satellite and with its own launch system.' Iran has sought a space capability partly because of America's growing regional presence. Developing these programs in response to the increased US presence indicates that Iran feels threatened and seeks to exploit space partly to safeguard its own national security (Kass 2006).

Iran apparently attempted to meet some of the above-noted goals starting from April 2003. The legislature approved a bill to create the Iranian Space Agency (ISA) to serve as a policy-formulating organization for space initiatives. The ISA carries out research on technology and remote sensing projects, develops national space equipment, and participates in the development of national and international space endeavors. It also coordinates various space-related activities within the country's research institutes, administrative agencies and universities. These efforts also help the ISA to execute decisions from the Supreme Aerospace Council.

It has been observed that Iranian efforts to advance its space program follow an unsettling pattern seen elsewhere. In slightly different ways, and with varying degrees of success, China, North Korea and Pakistan use a civil space program clandestinely to manufacture longer-range missiles to further safeguard national security. Iran seeks to become a space power for similar reasons (Kass 2006).

The Iranian Defense Ministry plays a prominent role in shaping the space effort, managing the Shahab ballistic missile program, which Iran modified to become a space launch vehicle (SLV). Testing a launch vehicle successfully has allowed Iran to boast that it is a space power (Kass 2006). In September 2000, an Iranian government spokesperson stated that the nation developed a modified missile, the Shahab-3D, to launch communication satellites. The Shahab-3D is a two-stage projectile that underwent a flight test in September 2000 using a combination of solid and liquid propellants. The Shahab system resembles North Korea's liquid propellant No-dong, which both countries agreed to develop jointly (Kass 2006).

The Iranian SLV initiative advanced further with North Korean assistance (Kass 2006). On 31 August 1998, North Korea attempted to launch a satellite by reengineering a ballistic missile. The Taepodong-1 failed to place its satellite into orbit because of a mechanical failure. None the less, the event marked an important advancement in North Korea's missile program. The country showcased some key requisites for developing longer-ranged missiles – multi-stage separation and advanced guidance mechanisms. Moreover, North Korea owned a multi-stage rocket capable of hitting targets much further than its more publicized cousin, the shorter-range and single-stage No-dong. Iran and Pakistan sent delegations to witness the 1998 launch. Their presence indicated that both nations could use Taepodong-1 technologies for their indigenous SLV efforts. Perhaps this event motivated Iran to conduct the September 2000 Shahab test using solid and liquid stages (Kass 2006).

On 27 October 2005, Iran met a key aerospace objective by becoming the forty-third nation to own a satellite. The Sinah-1 spacecraft entered orbit on board a Russian rocket, with the stated aim of monitoring natural disasters and observing agricultural trends. Moscow provided Iran with support. Sinah-1's primary mission was to demonstrate that Iran possesses an operational satellite.

Manufacturing an independent satellite is likely to occur through development of the 60 kilogram Mesbah spacecraft. The system is initially intended to obtain pictures for a variety of civilian purposes, to include greater data collection and distribution, assisting in efforts to find natural resources, and to predict the weather more accurately. Eventually, Iran will modify the satellite for remote sensing. The military could benefit from this technology, to look for suitable locations for building facilities. The Iran Telecommunications Research Center (ITRC) and the Iran Research Organization of Science and Technology (IROST) are jointly building this microsatellite with the Italian company Carlo Gavazzi Space (Kass 2006).

Construction of Mesbah began in 1997, just before the start of then Iranian President Muhammad Khatami's second term. On 4 August 2005, the day after Mahmoud Ahmadinejad succeeded Khatami, Tehran unveiled Mesbah. It was initially scheduled to enter orbit in early 2006 on board a Russian rocket but Iran may have also been exploring other options, to include the use of an indigenously-developed SLV, to launch Mesbah or other systems (Kass 2006).

There is widespread concern that Mesbah could serve as a springboard for Iran to manufacture more sophisticated reconnaissance satellites (Kass 2006). The Iranian Defense Ministry initiated an endeavor to manufacture the Sepehr satellite. Furthermore, Iran contracted with the Russian company, M. F. Reshetnev Scientific-Production Association of Applied Mechanics (NPO PM, Zheleznogorsk) to build the US$132 million Zohreh satellite. Zohreh is designed to provide Iranians with numerous services to include television and radio broadcasts, internet and e-mail access. Possessing advanced reconnaissance spacecraft could help Iran greatly, particularly after natural disasters, when emergency personnel could coordinate relief efforts more effectively. The military could also exploit this technology by rapidly distributing orders to forces to neutralize potential threats. Sending and receiving data quickly throughout the theater is a key characteristic of a sophisticated military presence, which Iran seeks to further modernize with space assets.

Iran fervently believes that it has a sovereign right to possess sophisticated technologies, including a space and nuclear program. Iran's return to space with an indigenously-produced SLV in 2009 made the country the first in the Islamic world with this capability (Kass 2006).

Saudi Arabia

The Kingdom of Saudi Arabia views space as a source of national pride (Kass 2006). It uses Russian technology to launch its spacecraft; the country's six satellites entered orbit on board Russian rockets. It has many SLVs that can carry small satellites, which is beneficial to the country because engineers have greater flexibility in the choice of a launch date. The engineers who built the SaudiSat-1A and SaudiSat-2 satellites acquired specialized knowledge, which will serve as a springboard for other related initiatives.

The spacecraft that Russia launched were constructed at the Space Research Institute (SRI). The SRI also supports the spread and advancement of space technology. The country's other notable space facility is the Saudi Center for Remote Sensing (SCRS), which was established in 1986, because Saudi Arabia recognized that remote sensing had numerous civil benefits. Enhancements to the SCRS allow it to obtain and distribute imagery simultaneously from multiple foreign remote sensing systems.

Saudi Arabia's space effort is far more mature than Iran's, yet generates significantly less international concern. According to Turki bin Sa'ud bin Muhammad al-Sa'ud, head of the King Abdullah University of Science and Technology (KAUST), his facility completely financed the SaudiSat-1A and SaudiSat-2 satellites without assistance from the Defense Ministry or any other government entity. The director stated that any claims that the two systems had a military purpose were 'baseless' – the satellites are intended solely for telecommunications and research purposes. KAUST's ability to fund this initiative completely demonstrates Saudi Arabia's desire to exploit space solely for civil purposes (Kass 2006).

Israel

Until recently, Israel enjoyed a regional monopoly in space (Kass 2006), being the only state in the Middle East region that could deploy satellites on board indigenously-manufactured SLVs. The Shavit SLV is a modified Jericho medium-range ballistic missile (MRBM) that placed numerous civilian and military systems in space. Hayim Eshed, head of the Israeli Defense Ministry Space Administration, boasted that 'With the exception of the Americans, we are superior to all other countries in two fields of satellite technology – resolution of photographs and picture quality.' Israel further enhanced its satellite technologies with the launch on 25 April 2006 of its Earth Remote Observation Satellite (EROS B) photo reconnaissance system. The system can also photograph activities at ballistic launch sites to obtain advanced warning of potential strikes and study upcoming missile tests. Israel has an edge over others in the region in satellite technology, which Iran seeks to overtake.

Israel officially entered the space age with the lift-off of its first satellite, Ofeq-1, from the locally-built Shavit launch vehicle on 19 September 1988. With that launch, Israel joined an exclusive club of countries that have developed, produced and launched their own satellites.

The next step was taken in early 2003, when NASA launched the twenty-eighth flight of the Space Shuttle Columbia, on mission STS-107. The seven crew members on board included the first Israeli astronaut, Ilan Ramon. The 16-day mission of Ramon and his colleagues was devoted to research, with over 80 experiments in Earth and space sciences, human physiology, fire suppression, and the effect of microgravity on a wide variety of natural phenomena.

The 1988 launch of Ofeq-1 was coordinated by the Israel Space Agency (ISA), established five years earlier to support and coordinate private and academic space-related research into areas such as electronics, computers, electro-optics and imaging techniques, which had already been in progress for some 20 years under the management of the National Committee for Space Research (Paikowsky 2007).

Designed as a technological satellite, Ofeq-1 spurred Israel's capability to send a satellite into orbit. Both Ofeq-1 and its successor, Ofeq-2, launched in April 1990, were very successful, sending back a stream of vital technical information. The two satellites reentered the Earth's atmosphere within six months of their launches. Ofeq-3 was launched in 1995 with an advanced electro-optical payload. It more than doubled its expected life-span, downloading images of superior quality. The unbroken success of Israel's satellite program was, however, brought to an abrupt halt with Ofeq-4. This fourth satellite in the series encountered problems in the second stage of its January 1998 launch; it burned up, setting back Israel's satellite reconnaissance program by several years. However, Ofeq-5 was launched successfully by a Shavit launcher in May 2002. Circling the Earth every hour and a half, Ofeq-5 is a reconnaissance satellite capable of delivering color images with an extraordinarily high resolution of less than a meter. Underlying the success of the Ofeq satellites and their comparatively inexpensive launch capability are Israeli developments in the field of miniaturization. Lighter satellites are more efficient and save hundreds of thousands of dollars per launch (Israel Minister of Foreign Affairs 2010).

Israel launched a microsatellite into orbit in June 1998. Developed at the Technion – Israel Institute of Technology – in Haifa for a mere US$3.5 million, TechSat II is a marvel of miniaturization. The satellite is an 18-inch cube that weighs just 106 lbs. It orbits 516 miles above the earth, generating its own energy from the sun, and is packed with miniature cameras, computers and other locally manufactured space hardware used in communications technology, remote sensing, astronomy and geoscience. TechSat II comes within photographing distance of Earth a dozen times a day. The ground-monitoring station at the Technion's Asher Space Research Institute downloads regular measurements of the atmosphere's ozone content from its ultraviolet sensors. From its charged-particle detector, scientists gauge the frequency with which such particles impact on the satellite and the potential damage they could cause to sensitive equipment such as computers. They also study the photographs recorded by its tiny camera (Paikowsky 2007).

Begun in the 1980s as a student project, TechSat rapidly extended its boundaries to become a professional satellite program. With the arrival of immigrant scientists from the former Soviet Union, the project took on its current form, making the Technion one of the few universities worldwide to have designed, built and launched a satellite.

As well as developing space hardware, Israel is using space as a platform to find out more about life on our own planet. In October 1996, ISA and NASA signed an active umbrella cooperation agreement, which allows Israeli life sciences experiments to be integrated with NASA space flights. The experiments conducted since 2005 have led to greater understanding in the fields of embryogenesis (the early development of mammals), osteoporosis (loss of bone density) and the setting up of 'space farms' to supply the spaceships and space stations of the future with food.

Many international space programs have taken an interest in Israel's space achievements (Paikowsky 2007). In addition to its links with NASA, Israel has formal space research cooperation agreements with France, Germany, Russia, Ukraine and the Netherlands. A similar agreement has also been signed recently with India, which provides for the installation of an Israeli-produced telescope on an Indian satellite due to be launched by 2011.

In June 2003, Israel was accepted into the European Space Agency (ESA) as a participating member. The agreement will allow Israel to participate in European space projects and to submit proposals for joint development projects.

The country sent its first geostationary telecommunications satellite into orbit on 16 May 1996. The Afro-Mediterranean Orbital System (AMOS) was built by Israel Aircraft Industries in partnership with Alcatel Espace of France and Daimler-Benz Aerospace of Germany. Launched by the French-built Ariane 4 launch vehicle, the AMOS communications satellite continues to provide high-quality broadcasting and communication services for the growing markets of Eastern Europe and the Middle East.

The TAUVEX (Tel Aviv University Ultra-Violet Explorer) scientific instrument constructed by El-Op (Electro Optical Industries Ltd.) is a vital component of a major international space research project in which Israel is an important player. The three-telescope array is designed to image astronomical objects in the ultraviolet range, including different types of hot stars (such as white dwarfs and mixed-type binaries), and young massive stars, which emit large amounts of ultraviolet radiation and ionize the interstellar medium, and are thus important in star formation processes and in the evolution of galaxies. TAUVEX operates in a spectral region with reduced sky background, thus can detect relatively faint objects with a modest observing time per target. TAUVEX was originally scheduled to be carried on the Indian satellite GSAT-4 as part of the India–Israel Agreement on Space Exploration. This multi-year mission could have yielded a very deep survey of parts of the sky, which might have enhanced considerably human knowledge of evolution in the Universe. Unfortunately, the ISRO decided in January 2010 to remove TAUVEX from the launch initiative.

A spin-off of TAUVEX is a small telescope with a resolution of 5 meters that will be used on the DAVID, a small, commercial remote sensing satellite. Developed jointly by an Israeli hi-tech company and a German firm, the project is supported by the EU and ISA. Israel will participate in the European Global Navigation Overlay System (EGNOS), as well as the new Galileo project.

Pakistan

The Space and Upper Atmosphere Research Commission (SUPARCO) was established by Pakistan in 1961, for the purpose of space science research and development, and began operations in 1964. This national organization, with a high degree of autonomy, implements the space policy of Pakistan and was established by the Space Research Council (SRC), whose president is the prime minister. The commission comprises the chairman and four members for space technology, space research, space electronics, and finance, respectively.

It was granted the status of a Commission in 1981. SUPARCO is devoted to R&D work in Space Sciences and Space Technology and their applications for the peaceful uses of space. It works towards developing indigenous capabilities in space technology and to promote space applications to boost the socio-economic status of the country (Lele 2005).

Badr-1 was Pakistan's first indigenously developed satellite. It was launched from the Xichang Launch Center, China on 16 July 1990 aboard a Chinese Long March 2E rocket. Badr-1 weighed 150 pounds. Originally designed for a circular orbit at 250–300 miles altitude, Badr-1 was inserted by the Long March rocket into an elliptical orbit of 127–615 miles. The satellite successfully completed its designed life.

SUPARCO started building the small amateur radio satellite in late 1986, with support from the Pakistan Amateur Radio Society. The satellite was named Badr, from the Urdu word for 'new moon'. It was planned that Badr-1 would be launched on the US Space Shuttle, but the 1986 Challenger explosion and consequent delay in American flights meant a change of plan (Lele 2005).

Paksat-1 was Pakistan's first geostationary satellite. The satellite was originally known as Palapa-C1 and was designed to serve Indonesia. After an electronics failure, it was renamed Anatolia-1 and then renamed again as Paksat-1 in 2002. It was originally manufactured by Boeing and used the HS 601 spacecraft design. It was launched on 1 February, 1996.

The Paksat-1R satellite will replace the existing Paksat-1 in 2011. Pakistan's national space agency signed a consulting deal with Telesat in March 2007 regarding advice on the purchase, manufacture and launch of the Paksat-1R satellite. Under this agreement, Telesat will help the Pakistani agency to find a manufacturer, and provide technical and commercial advice during the negotiation process. Telesat will also help to oversee the construction of the new satellite, and monitor the launch and in-orbit testing services (Lele 2005).

Japan

The Japanese equivalent of NASA is the Japan Aerospace Exploration Agency (JAXA). Japan's space development competitiveness has been ranked seventh in the world behind India and Canada. Its annual space budget of US$2.5 billion is one-fourteenth of that of the USA and one-half of that of Europe. Japan launched its first mini two-stage rocket in 1957. This launch is regarded as the beginning of the Japanese space program. Japan has two launch centers in Kyushu: one near Kagoshima and the other on the southern island of Tanegashima (JAXA 2010).

The Japanese space program is helped by access to advanced technology produced at home and friendly relations with the

USA, but is restricted by tight budgets and limited popular support. The program has gotten a boost as result of competition from China and India – which have been making great advances is space even though they lack Japan's advanced technology (JAXA 2010).

Two Japanese spacecraft swept up material in the original cloud of dust and gas from Halley's Comet in the 1980s. In November 2005, a mini-robot vehicle designed to sample rocks was sent to investigate the surface of an asteroid but was lost in space for several years after it was released from a spacecraft only 55 meters from the asteroid. It was located again in 2010. The vehicle was only 10 centimeters long and weighed 600 grams.

In October 2008, the Japan Origami Association announced that it had created a paper airplane made with a special heat-resistant, glass-coated paper that its makers said would allow it to survive reentry into the atmosphere from space and float to earth. A prototype passed a durability test in a wind tunnel in March 2008. The makers hope to release the plane from the International Space Station.

Japan has very well developed rocket production facilities (JAXA 2010). The Japanese-built H-2 rocket is 48 meters in height, weighs 260 tons and can carry a 2,200 kg payload into geostationary orbit. It had five successful launches before 1998 but was very expensive – costing Japan twice as much to launch per payload as Europe's Ariane rockets – partly because of Japan's insistence on the rocket being totally made in Japan. The M-5 is a three-stage solid-fuel rocket capable of lifting 1.8 metric tons into orbit at a height of 250 km. Six were launched. All but one launch was successful, but the rockets were scrapped because they were too expensive (each launch cost ¥7 billion). The H-2A is a 290-ton, 53-meter rocket first launched on 29 August 2001. Made with foreign parts that reduced its costs by half, it has boosters and two liquid-filled engines, and has been used to launch commercial and spy satellites. In the future it hoped that the H-2A will be used to launch an unmanned space shuttle. Beginning in 1998, three launches – two H-2s and an M-5 –ended in failure, costing Japan 18 launch contracts with two US satellite manufacturers. A launch failure of an H-2A in 2003 was described as 'the world's most expensive firework's display'. Japan had six successful launches in 2006, four with H-2A rockets. In December 2006, the country successfully launched its heaviest satellite ever, the Kiku No. 8 test satellite. Weighing 5.8 metric tons, it was launched from an H-2A liquid fuel rocket with four solid-fuel boosters. The satellite was placed in geostationary orbit and uses two 18-meter-long antennae. There were some problems with the unfolding of the antennae. As of November 2009, there had been 10 consecutive successful H-2A launches and 15 successful launches overall. After the unsuccessful launch in 2003, the H-2A had eight successful launches. As of 2008, the success rate of the H2-A rocket is 93 percent (11 out of 12 launches), which is deemed good enough for commercial applications. In January 2010, the IHI Corp decided to liquidate the GX rocket development program after the government pulled out of the project the previous month. The GX rocket project was scrapped because it cost too much money and its prospects as a commercial satellite launcher seemed dim (JAXA 2010).

As for satellites, the first Japanese satellite, the 24-kg Ohsumi, was launched in 1970. In March 2006, Japan successfully launched its first infrared satellite, the Astro-5 observation satellite, with an infrared telescope designed to search deep space for planets. It was launched from Uchinoura Space Center on a three-stage, 30.8-meter-high M-5 rocket. It was the third successful space launch in a month. Two previous satellites had been sent up on H-2A liquid fuel rockets. In September 2006, Japan successfully launched a solar observation satellite on an M-5 solid fuel rocket. The third solar observation satellite launched by Japan, it is equipped with telescopes capable of detecting optical wavelengths, X-rays and extreme ultraviolet waves. In February 2008, Japan successfully launched a 2.7 metric ton satellite, designed for high-speed internet transmissions, on an H-2A rocket. The project was delayed and cost US$500 million, and many questioned its cost-effectiveness, especially when considering that land-based internet service has become much faster and more efficient. Japan had hoped to set up its own version of a GPS but that project looks as though it will collapse because of a lack of a real need for such a system.

In January 2009, eight satellites were launched successfully on board an H-2A rocket. The satellites included one that will observe lightning and carbon dioxide concentrations.

In March 2003, Japan launched its first two spy satellites from Tanegashima Island off Kyushu. They were launched with an H-2A rocket and placed in polar orbit at altitudes of 400 to 600 km. One of their main purposes is to keep an eye on North Korea and its nuclear reactors and missiles. The satellites were made by a consortium of Japanese companies headed by Mitsubishi Electric and are part of a US$2.1 billion public and private project (JAXA 2010). After China successfully blew up a weather satellite in space in January 2007, Japan quietly overturned a law restricting its space program from being put to use for military purposes. Japan has launched its own spy satellites because it does not want to rely on American satellites. Images from American satellites are often expensive and delayed, and often requests were turned down for security reasons. With its own satellites, Japan can get information whenever it wants. The launch of a second set of spy satellites in November 2003 was a failure. The satellites and rockets were destroyed because one of the two boosters failed to separate. In February 2007, Japan put its fourth spy satellite into orbit, giving its satellite system full global coverage and allowing the Japanese military to survey the entire planet: and to photograph any point on Earth at least once a day. Of the four satellites, two have optical sensors and two have synthetic aperture radar. The optical satellites can discern objects as small as one meter on the ground. Each of these four satellites is designed to orbit the Earth 15 times a day, and observe any location on Earth at least once in 24 hours. Within a few years Japan hopes to have 16–20 spy satellites orbiting the Earth.

Japan had hoped to enter the commercial satellite business in the 1980s (JAXA 2010), and a good deal of government money was pumped into rocket and satellite projects with that goal in mind. The current cost of launching a satellite with an H-2A rocket is between ¥10 billion and ¥12 billion. Analysts say it is necessary to reduce the cost to ¥8 billion to compete with Europe and the USA. In 2007, the H-2A rocket program was privatized and transferred from JAXA and the government to Mitsubishi Heavy Industries (MHI). In April 2007, MHI announced that was in discussion with European competitors to work together to offer satellite-launching services. The satellite-launching market has been dominated by Russia, Europe and the USA. Japanese launches cost about 20 percent to 30 percent more than European ones, and the launch period is limited to 190 days a year because of concerns about damaging fishing boats in the launch area. Japan aims to become more competitive in the satellite industry by developing mini satellites that will be considerably cheaper to launch because of their size. The aim is to develop satellites with desired features that weigh between 100 kg and 300 kg rather than the standard 3 metric tons, and launch them by 2011. In October 2008, MHI announced that it was in the running to obtain a contract to launch a South Korean weather observation satellite in an H-2A rocket. The company's bid was successful, and the launch of the South Korean reconnaissance satellite Kompsat 3 is scheduled for 2012. This will be the first commercial launch on an H-2A rocket (JAXA 2010).

As for space exploration, several Japanese people, including one woman, have gone on Space Shuttle missions. Some have brought along special spherical 'noodles' that are easier to swallow in zero-gravity than conventionally-shaped noodles. Japanese astronaut Soichi Noguchi was on the Space Shuttle Discovery flight in July 2005, the first shuttle flight for over two years. Noguchi participated in space walks during which repairs were made to tiles damaged during take-off. In March 2008, Japanese astronaut Takao Doi went into space aboard the Space Shuttle Endeavour and became the first Japanese to participate in a docking procedure when the space shuttle docked with the International Space Station (ISS). Doi's primary mission was to install the 4-meter-long experiment module for Japanese Kibo laboratory on to the ISS using the shuttle's robotic arm. In June 2008, on another space shuttle mission, the Kibo storage facility was installed on the experiment module. Japanese astronaut Akihiko Hoshide was responsible for operating the bolts to join the sections in the early and latter stages of the process. Daisuke Enomoto, a former executive with the troubled internet company Livedoor, paid a US\$21 million space tourist fee to fly on a 10-day trip to the ISS aboard the Russian

Soyuz capsule in September 2006, but was pulled from the three-person crew at the last minute and replaced with Dallas businesswoman Anousheh Ansari, reportedly because he had kidney stones. Space Adventures, the space tourism company that arranges trips into space, said the cancellation of the flight for medical reasons meant that Enomoto was not entitled to a refund of his US$21 million. Enomoto sued Space Adventures, saying the company was well aware he had kidney stones when he signed up for the trip, and that Ansari was given his place because she had invested in Space Adventures. Japanese astronaut Koichi Wakata was selected as the first Japanese astronaut to undertake an extended stay on the ISS. He was delivered to the space station by the Space Shuttle Discovery in March 2008 and stayed on it for three months. Some of his chores were to operate the robotic arm, which he helped to design; to check the Discovery for tile damage; and to install solar wind panels on the ISS. Other goals included cultivating cells that would develop into frog kidneys, and monitoring how his body reacted to being weightless for extended periods of time. In December 2009, Japanese astronaut Soichi Noguchi arrived at the ISS for a 6-month stay, during which time he was involved in scientific studies making use of the space environment; growing a space garden of mint, dandelions and other plants; and engaging in the installation of the robotic arm on the Kibo laboratory. Noguchi's 6-month stay is the longest space trip among Japanese astronauts; 5 months was the previous record. It was Noguchi's second trip in space. Because he stayed for such a long time he was given a private room in the ISS. On 5 April 2010, Naoko Yamazaki was launched into space on the shuttle Discovery as part of mission STS-131, returning to Earth on 20 April 2010 She was Japan's eighth astronaut and only the second Japanese woman to go into space.

Kibo, which means 'hope' in Japanese, is Japan's first manned space facility (JAXA 2010). An attachment to the International Space Station, it has room for up to four astronauts to perform experiments for a long period of time. The project was launched in 1985 and suffered delays, overruns and near cancellation under the US Clinton Administration, to finally make it to the ISS in 2008, 15 years behind its original schedule. Among the experiments conducted on Kibo are ones controlling the crystallization process of silicon melting that can lead to the production of stable, homogeneous chip materials, which are known to be susceptible to liquid convection. The Kibo laboratory was completed in July 2009 at a cost to Japan of $7.6 billion.

In September 2009, Japan successfully launched the H-2 Transfer Vehicle (HTV), Japan's first unmanned spacecraft, aboard an H-2 rocket and maneuvered it to dock with the ISS. Japan hopes the HTV will be the main supply vehicle for the ISS after the Space Shuttle is retired. The development of the HTV began in 1996. Perhaps its most remarkable feature is its ability to be maneuvered to another spacecraft and to dock with it using remote control. On its maiden voyage it used the global positioning system to maneuver from 5 km from the ISS to within 500 m of it. From there it was gradually and carefully eased to the ISS by controllers on the ground. To prevent a collision, the HTV was not allowed to proceed until it had cleared several checks along the way. Crew members aboard the ISS had the power to stop the HTV if necessary. The ISS moves at 7.7 km per second, 350 km above the Earth. Successfully docking the HTV with the ISS, as one JAXA official said, 'is as difficult as threading the eye of a needle via remote control'. One of the HTV greatest attributes is that it is unmanned. If something happens to it there is no danger of loss of life, and expensive safety and back-up systems used for astronauts do not have to be installed (JAXA 2010).

In September 2007, Japan launched its first probe to the moon. The probe, the Kaguya lunar explorer, consisting of a main satellite and two sub-satellites, took extraordinary high-definition pictures of the dark side of the moon and a full image of the Earth from above the moon as well as scanning the surface with X-ray and infrared devices from between 100 km and 800 km above the lunar surface. The probe was launched on H-2A rockets built by the private company, Mitsubishi Heavy Industries. The cost of the project was around US$500 million. The launch of the Japanese moon probe came a week before China launched its own lunar probe and was seen as an escalation of Asia's undeclared space race, which also includes India, who launched a lunar probe in April 2008. Japan has plans for a spacecraft to land on the moon in the 2010s and release a lunar rover similar to the ones used by the USA on Mars. One plan calls for a Japanese manned mission to the moon between 2025 and 2030. Japan has hopes of mining materials for fusion reactors on the moon and setting up an observatory there. Shimizu Corp., a large Japanese construction company, has a team researching the construction of a base on the moon. It has developed technology for manufacturing concrete on the lunar surface and building solar-powered satellite power plant. Japan is experimenting with prototypes of a reusable space capsule and a delta-wing space shuttle. Japanese scientists are working on a solar sail to be used to propel spaceships using sunlight. In 2010, Japan planned to put into space a 2.1-meter-long probe to Venus to study the climate there. Japan and Russia have discussed conducting joint missions to Mars that would involve sending a probe to one of the Martian moons to collect soil and bring it back to Earth in what would be the first round-trip journey to Mars or one its moons. A Japanese mission to put a satellite in orbit around Mars came close to succeeding (JAXA 2010). Nozomi (Planet-B) was Japan's first Mars satellite explorer and its main mission was to research the Martian upper atmosphere by focusing on interaction with the solar wind. The satellite was launched on 4 July 1998 from the Kagoshima Space Center in Uchinoura. On the way to Mars, however, problems arose and substantial orbit changes had to be made. It finally came close to Mars in December 2003, four years behind the original schedule. However, there were further problems and the systems required to enter orbit around Mars failed despite every effort to restore them, and on 9 December 2003 JAXA was forced to abandon the placing of the satellite into the chosen orbit. Nozomi thus became an artificial planet flying for ever around the Sun near the orbit of Mars.

4

The future

Scenarios for the future of the space sector

To explore the future, analysts can choose among various techniques, depending on the nature of the exercise involved. Forecasting is perhaps the most prevalent technique. It employs forecasting models that provide a simplified description of reality and of the relations that are believed to exist between independent or exogenous variables (the values of which are determined outside the model) and dependent or endogenous variables (the values of which are generated by the model). Forecasting models are useful for short-term projections, but they are of little value for exploring the long-term future. This is because such models implicitly assume that the underlying structure of the model (more specifically the relation between the dependent and independent variables) does not vary over the forecasting period.

While this assumption may be reasonable for the short term, it is unlikely to hold in the longer term. Attempts can be made to deal with this problem by developing several forecasts based on alternative values of some of the structural parameters. However, in this approach, uncertainty is treated as an excursion around a 'preferred' or 'most likely' path or destination. For futures that are inherently unpredictable, a range of scenarios offers a superior alternative for decision-making, contingency planning or mere exploration, since uncertainty is an essential feature of scenario analysis. Individual scenarios provide a rich characterization of alternative futures. Their goal is to describe a coherent future world by means of a credible narrative. Taken together, several scenarios are likely to contain the future state, though no individual scenario would be able to describe it. Here, the scenario approach is clearly preferable, since the drivers are broadly defined and involve complex interactions with a wide range of variables over a long period of time.

This section analyzes three alternative future visions of the world, and their implications for the evolution of the space sector. For each likely scenario, the political, economic, social, energy, environment and technology aspects are presented, and the consequences for the space sector (considering military, civil and commercial components) are predicted. These forecasts are based on the main trends and factors likely to influence the drivers of change up to the 2030s, according to opinions expressed by experts in the recent literature and collected in the OECD's (2004) publication *Space 2030: Exploring the Future of Space Application*.

Scenario 1 – 'Smooth sailing'

This is an optimistic scenario where the world is at peace, multilateralism and international cooperation prevail, and globalization brings prosperity to the world, notably the developing world. Poverty is significantly reduced, energy supplies are adequate to meet demand and effective measures to clean up the environment are taken collectively. There is a pacific global world order, guided by international organizations, in which free markets and democracy gradually become the universal model for social organization. International co-operation allows for the rapid growth of the space industry.

Main features
- Political: a strong spirit of cooperation. The USA, the European Union, Japan, Russia, China and India have good relations, and are more interdependent. However, they still face the threat of the use of weapons of mass effect by terrorists and criminal groups.
- Economic: strengthening of the World Trade Organization (WTO). Foreign direct investment (FDI) is better protected.

High rates of growth worldwide, as developing countries gradually catch up with the West. Demand for transport and communication, and educational services, increases rapidly.

- Social: A growth in prosperity helps to lighten the adverse consequences of demographic trends. In developed economies, economic growth provides the money to deal with the costs of an aging population. In developing countries, it generates jobs for the rapidly growing labor force. As a result, migration flows from the South to the North increase moderately. More effective public health and education programs are implemented worldwide.
- Energy: international tensions over energy remain, as alternative sources of energy are developed. It is not easy to meet rising demand, and big efforts must be made in exploration and extraction.
- Environment: environmental problems tend to increase in the medium term. The EU takes the lead in seeking solutions, and a new world treaty substituting the Kyoto protocol is put in place.
- Technology: new advances spur economic growth and help to fight environmental threats. Fast diffusion of new technologies to the developing world helps these countries to catch up with the West. Technology is developed increasingly at the international level, as cooperation among national research teams is becoming tighter and tighter.

Implications for the space industry

Major progress is achieved in applying space technology to the solution of global social and environmental problems.

In the field of military space, a more peaceful world order puts less priority on military expenditure. As tensions among the space powers are reduced, cooperation among them increases. The focus is put on military space infrastructure in the areas of telecommunications, Earth observation and navigation.

In the field of civil space, space exploration and investments in science will increase significantly. An increasing number of countries decide to join the International Space Station (ISS). By 2020, a permanent international station is established on the moon; by 2025, the first manned mission to Mars is launched. In terms of the development of civil space infrastructure, the International Space Agency (ISA) will support international efforts toward space-based solutions to global problems. The World Health Organization (WHO) will support the use of tele-medicine in the developing world; and UNESCO will promote distance learning as a way to reduce educational backwardness. Space assets will be used for monitoring crops, pest control and precision farming. Private Western firms will participate in these programs. A world environment protection agency will be created, which will use space-based resources for monitoring the enforcement of environmental agreements.

In the field of commercial space, we shall witness the creation of an even more open business environment. WTO discipline will be extended and space firms will benefit from FDI protection. Similarly, UN conventions on space will provide clearer definitions and better reflect commercial property rights. The role of the International Telecommunications Union (ITU) in allocating orbital slots will be enforced. All space nations will adopt space legislation following a uniform model code. Regulations regarding operators of space assets will be harmonized. There will be a significant expansion of the space infrastructure as the result of the establishment of a global broadband telecommunications infrastructure as well as truly global positioning and navigation infrastructure for civil and commercial use. Global Earth observation system will be used for civil security and commercial purposes. In terms of development of the space industry, space firms will be able to restructure globally to take full advantage of economies of scale and scope. Space firms will engage in fierce competition. Major efforts will be made to cut costs appreciably and improve services. Large R&D budgets will be devoted to innovative space products and services. The cost of accessing space will be reduced significantly. The cost of manufacturing launchers will decrease substantially, and major advances will be made in micro- and nanosatellites. Space tourism will begin, first on a sub-orbital basis, and then on an orbital basis by the 2020s.

Scenario 2 – 'Back to the future'

This is a 'middle of the road' scenario that describes a return to a bipolar world where international relations are dominated by the uneasy interaction between two blocs: the USA and Europe, on the one hand, and a coalition between China and Russia on the other. Despite the political difficulties, economic growth remains reasonable under an economic regionalization scenario involving closer cooperation between the USA and Europe. However, tensions are on the rise on several fronts, notably over the environment and energy security. According to the scenario, as in the Cold War era, a bipolar world gradually emerges. US dominance is challenged by a growing China, which rejects Western values and is supported by a recovered Russia. Europe strengthens its ties with the USA and coordinates its military forces. Conflicts between the two blocs are constants.

Main features

- Political: China is keen to take advantage of its new strength but its national politics cause conflict with the Western countries, mainly over energy and natural resources. China is helped by Russia, because the two economies are becoming complementary. Both blocs respond to tensions by strengthening their military capability.
- Economic: poor economic growth is achieved in the West, as the USA cannot improve its trade deficit and Europe cannot undertake the necessary structural reforms to spur growth. In contrast, China enjoys high rates of growth; but its attempts to satisfy its huge demand for supplies result in confrontation with the West, which responds by closing markets to Chinese products, thus dividing the world economy into two rival blocs.
- Social: as economic growth is poor, social tensions emerge. Immigrants are viewed with hostility; more emphasis is placed on order and security in aging societies.
- Energy: heavy dependence on fossil fuels continues. Concerns about the security of supplies rise, and major efforts are put into developing new sources of energy.

- Environment: international agreements disappear, as the environment deteriorates. Agreements on controlling pollution are only at the regional level.
- Technology: the rate of innovation in the West is affected by slow economic growth. Priority is given to military research.

Implications for the space industry

Three cooperative blocs emerge: USA–Europe–Japan, China–Russia, and India–other emerging countries (for example, Brazil). Space firms benefit from greater military effort, but also suffer from a less open trade and investment environment.

In the field of military space, a new type of space race and the 'weaponization' of space spur the competition among blocs about the military use of space technology. Budgets are spent mainly in developing anti-satellite systems and space-based lasers, which are capable of attacking both satellites and missiles. The military space industry of the USA and Europe is integrated, and so is that of China and Russia.

In the field of civil space, main civil space projects are devoted to creating 'soft power': prestige and influence. In particular, in terms of space exploration and science, the Western bloc launches an unmanned Mars exploration program, with the aim of landing humans on Mars by mid-century. China and Russia start a moon project, with the aim being to establish a manned outpost and exploit the moon's potential mineral and energy resources. India promotes cooperative space projects among developing countries such as Indonesia and Brazil, thus establishing a third axis in the new space race. As for the development of civil space infrastructure, space applications increase and provide government-sponsored solutions. Telemedicine and efforts focused on the environment are the main projects. Private actors develop sub-orbital launchers, spurring space tourism. The main infrastructures are devoted to military uses: energy relay satellites, space-plane technologies and so on.

In the field of commercial space, a return to protectionism in the space sector is encouraged by security concerns. In terms of the business environment, the emphasis on military applications tends to slow the development of commercial space, as space firms devote a higher proportion of their resources to military contracts. Many new space-related products are developed regionally, but trade restrictions reduce the diffusion of new technologies. Restrictions on information flows have a negative effect on the telecommunications sector. The growing demand for energy results in further exploration and a greater need for space-based technologies (to monitor pipelines or to improve exploratory techniques). Sub-orbital space tourism is developed to a certain extent by private firms. Semi-private firms integrate their activities and develop dual-use applications under public–private partnerships, taking advantage of the military efforts.

Scenario 3 – 'Stormy weather'

This relatively pessimistic scenario describes a world where a breakdown in multilateralism, caused by a strong divergence of views among key actors, precipitates an economic crisis that further exacerbates international relations. Economic growth is likely to be slow, and concern about the environment is low. The worst nightmares come true. International institutions are gradually eroded and ignored. International cooperation is replaced by bilateralism, as ethnic conflicts multiply, leading to massive migrations and terrorism. Economic conditions deteriorate as the world returns to protectionism.

Main features

- Political: confronted with terrorism and other threats, the USA gradually becomes isolationist. Tensions between the USA and other countries, including its European allies, are frequent. A confusing web of shifting partial agreements/alliances among like-minded countries emerges.
- Economic: slower growth and gradual erosion of the WTO discipline. Tariff and non-tariff barriers emerge, and flows of FDI dry up, as the globalization process is gradually reversed.
- Social: security concerns move to the top of the policy agenda. Poverty rises in the South, as migration flows to the North increase dramatically.

- Energy: security of supply is the primary concern for most countries, exacerbating tensions among importing countries.
- Environment: is not paramount in the policy agenda, as security and energy take all the attention. Anti-pollution measures are put in place at a national level, as international agreements are difficult to reach.
- Technology: depressed economies and lack of cooperation bring about a low rate of innovation, except in the military sector.

Implications for the space industry

Security and defense uses of space become paramount in a divided world with no clear alliances. The impact on space business is mixed: on the one hand, space firms benefit from government contracts; while on the other, markets become more fragmented.

In the field of military space, the budget for this rises worldwide. The USA, which maintains its lead in the space industry, develops an unmanned reusable hypersonic cruise vehicle for military purposes. Europe launches a major military space program in the 2010s, in order not to be left behind in military capability by the USA, and emerging countries such as China. China begins a similar program.

In the field of civil space, space exploration and science, no major common international exploration programs are pursued. Some countries try to strengthen their soft power through a number of spectacular initiatives. However, the scientific value of these space ventures is weakened by duplication of effort and by the priority given to technology over science. In terms of civil space infrastructure, China and India are the leaders in the development of space-based telemedicine and distance education applications, ready to be exported to other developing countries.

In the field of commercial space, given that the business environment is one in which protectionism tends to be strong, this limits technology transfers and export opportunities. Private investment in space is reduced, because of the depressed economic conditions. This is partly counterbalanced by the decision of a number of governments to purchase space services directly from private sources, rather than to create them themselves. There will be a limited expansion of the commercial space infrastructure as a result of strong regional barriers to information exchange, which will have a damaging impact on telecommunications. Space assets will be used for monitoring the production and distribution of oil and gas. A civil and commercial version of the small launch vehicle will give the USA an advantage in launching small satellites. Sub-orbital space tourism will hardly develop any further.

Forecasts for space applications

Having defined the likely scenarios for the future of the world in general, and the space industry in particular, the OECD report (OECD 2004) focused on concrete space applications and identified potentially promising applications that are likely to be both in demand and technically feasible in the coming years. To this end, the report identified the space value chain as a set of three broad groups of activities or services: information services, transport services, and manufacturing. Information services are likely to be developed first, given the high cost of access to space. Transport applications would follow, as they rely on information applications. Manufacturing/mining applications, which depend on the effective development of the other two groups, would be expected to be the last of the three. Under the conditions of the future scenarios presented in the previous section, demand and feasibility for the three broad groups of services are analyzed below.

Potential future demand for information applications

Telecommunications

Table 4.1 shows that potential demand for telecommunications remains strong, and focused on the broadband (fourth generation).

Table 4.2 shows the potential growth of commercial and social demand for specific telecommunications services. Demand changes significantly in the three scenarios. Demand for mobile

Table 4.1 Summary of expected demand for telecommunications

Scenario 1 Smooth sailing	Scenario 2 Back to the future	Scenario 3 Stormy weather
Given increasing liberalization of the economy, significant expansion of demand for communications, especially for broadband at both domestic and international levels.	Less expansion than in Scenario 1 because of constraints imposed by the bipolar world, but large within major regions.	International demand less than in Scenario 2, focused on broadband.
Space is in a good position to capture a sizeable share of demand because of its ubiquity as an integral part of a global broadband network.	Role for space less important than in Scenario 1 because of the perception that space assets may be more vulnerable.	Smaller role for space than in Scenario 2.

Source: OECD (2004).

Table 4.2 Growth of potential demand for space-based information services

	Scenario 1 Smooth sailing	Scenario 2 Back to the future	Scenario 3 Stormy weather
Multimedia entertainment	High	Medium	Medium
International e-commerce	High	Medium	Low
Distance learning	High	Medium	Low
Telemedicine	High	Medium	Low

Source: OECD (2004).

communications should be high. Space telecommunications face serious terrestrial competitors. In conclusion, given the size of the untapped potential markets for telecommunications, the main promising applications are: telemedicine, distance learning, e-commerce and multimedia entertainment.

Earth observation

Earth observation (EO) is a space application that is technologically mature and very valuable for military, civil and commercial purposes. From a military perspective, EO is a critical component of intelligence, communications, command and control. From a civil perspective, EO supports important public responsibilities, including security, the management of natural resources, urban planning, weather forecasting and climate change monitoring. From a commercial perspective, EO covers a growing range of businesses, from insurance companies to farmers. Tables 4.3 and 4.4 show that demand is expected to increase in all scenarios, though the composition of demand varies.

Table 4.3 Summary of expected demand for Earth observation

Scenario 1 Smooth sailing	Scenario 2 Back to the future	Scenario 3 Stormy weather
Demand expected to be strong for civil, security and commercial applications across a broad range of activities. Systems likely to be regional or global and fully integrated.	Demand expected to be strong for military, security, civil and commercial purposes. Systems likely to be regional.	Demand expected to be strong for military and security; less for civil and commercial. Systems largely to be regional.
Key role for space in association with other techniques. Space competitive because duplication of effort is minimal.	Important role for space, perhaps not as effective as in Scenario 1 because of duplication of effort.	High military demand; high cost for civil and commercial.

Source: OECD (2004).

According to Table 4.4, excluding military applications, the most promising applications in EO would be: environment applications (meteorology, climate change); land use management (urban planning, precision farming); exploration (oil, gas); natural disaster prevention and treaty monitoring.

Table 4.4 Potential demand for space-based Earth observation services

	Scenario 1 Smooth sailing	Scenario 2 Back to the future	Scenario 3 Stormy weather
Meteorology	High	High	High
Precision farming	High	High	Medium
Fisheries	High	Medium	Medium
Forestry management	High	Medium	Medium
Exploration (oil, gas)	High	High	High
Urban planning	High	High	High
Natural disaster prevention	High	High	Medium
Defense/security	Medium	High	High
Treaty monitoring (environment, disbarment)	High	Medium	Medium

Source: OECD (2004).

Potential future demand for space transport and manufacturing

Space tourism

As Table 4.5 shows, in all three scenarios there is a strong drive to reduce the cost of access to space, notably by developing a genuinely reusable launch vehicle. However, the growth of space tourism depends on international cooperation over civil goals, and this is achieved only in the first scenario. In conclusion, based on social and commercial demand but highly dependent on the development of space transport and security, sub-orbital tourism and orbital space tourism are viewed as a limited application.

Table 4.5   Summary of expected demand for space tourism

Scenario 1 Smooth sailing	Scenario 2 Back to the future	Scenario 3 Stormy weather
As one of the largest and fastest-growing industries, tourism expands significantly despite concerns about security.	Growth of tourism industry less rapid than in Scenario 1.	Growth of tourism industry slower than in Scenario 2.
Some tourist attracted by a space adventure if available at a price they can afford.	Fewer potential candidates than in Scenario 1.	Fewer candidates than in Scenario 2.

Source: OECD (2004).

Space production activities

Space production includes three types of activities:

- In-orbit manufacturing: for example, manufacturing and testing of pharmaceutical products in microgravity. It may emerge for very highly-value items (new alloys, composites) if the cost of access to space is significantly reduced.
- Space power generation: for example, development of space solar power systems to provide energy from space to Earth. The economic potential exists, but the ability to produce energy in space and transmit it to users on Earth is far from technically feasible at the time of writing.
- Extraterrestrial mining: to provide new resources for Earth or to build outposts. It is an activity for which demand is not well defined. Technical hurdles are important now.

According to the report (OECD 2004), the overall development of the space production sector will depend on a drastic reduction of the cost of access to space, and on the availability of cheap sources of energy in space as well as on the evolution of space production processes and techniques.

Technological feasibility for space applications

Whether a particular space application has a chance to flourish depends not only on whether it is likely to be in demand, but also on it already being, or is expected to be, technically feasible, and whether it may be available at a price that will make it attractive to users. An analysis of the 'enabling technologies' (technologies that are expected to facilitate the implementation of the application and to reduce its cost) should be made. The enabling technologies considered for the space sector are nanotechnology; biotechnology; information and communication technologies (ICT); manufacturing technologies; and robotics and artificial intelligence (AI). Many of the potentially promising applications we have identified previously are not only technically feasible at the time of writing, but are likely to become even more attractive in the future (for example, telecommunications, Earth observation). But other applications need the development of interlinked technologies before they can be considered feasible. As far as future feasibility is concerned, space applications can be divided into two groups, namely main contenders and outsiders. The main contenders are those applications that appear to have a reasonable chance of

Table 4.6   Ranking of feasibility of promising applications

Main contenders	Outsiders
Distance learning and telemedicine	Adventure space tourism
E-commerce	(sub-orbital, orbital)[1]
Entertainment	In-orbit services
Location-based consumer services	Power relay satellites
Location-based services: traffic management	
Land cover: precision farming	
Land cover: urban planning	
Land cover: exploration (oil, gas)	
Disaster prevention	
Meteorology and climate change	
Monitoring: treaties, policies	

Note: [1] Although there has been significant progress since 2004.
Source: OECD (2004).

flourishing, while outsiders are those applications that have less of a chance of realization over the upcoming 30 years, even though demand conditions appear to be favorable. Table 4.6 summarizes these applications.

Comparison with other industry sectors: from aviation to space tourism

Another way to forecast the future of the space industry is to look at the evolutionary patterns of more mature industries, such as the aviation industry. Following a description of the parallels between the two industries, we hope to be able to offer some more accurate projections for the future of the commercial space industry, both in terms of an overall strategy internationally and of the future of the industry in general.

Aviation industry

There are some similarities between the evolutionary processes of the space industry and the aviation industry; we have outlined some of these similarities below.

Stage I: Military beginnings

Aviation industry

The First World War was the catalyst that prompted aviation industry advancement (Pritchard and MacPherson 2005). Two cycles emerged which, in steady rotation, promoted the future of aviation. One cycle was the 'parry and thrust' between adversaries – as one side gained a competitive advantage, the other required better planes to counter that advantage. The other cycle was a similar form of competition between the makers of fighters and bombers – the better a fighter, the better a bomber was required to be, to offset the adversary's gain, and vice versa. Military aviation began with the use of reconnaissance air-craft – used to establish the position and activities of the enemy. The French army purchased its first reconnaissance planes in

1910 and one year later introduced armament into these craft. In 1912, the French military began experimenting with aerial bombing, followed by the British in 1913. Meanwhile, Igor Sikorsky built the 'air giant' in Russia, a four-engined aircraft that became the prototype of the multi-engine strategic bombers of the First World War.

Space industry

The space industry, like the aviation industry before it, has its roots in military motivation. The Cold War, which lasted from the mid-1940s to the early 1990s, spurred the 'space race', which involved informal competition between the USSR and the USA to be the first to exploit space exploration and technology for military defense purposes. The term 'space race' in fact emerged as an analogy to the 'arms race': 'From the start, the Space Race was an Arms Race' (Broad 2007). As a result, the Cold War period became an era of outstanding success for emerging space programs (Hughes 1990). The Soviet Union's launch of Sputnik 1 in 1957 was the catalyst that spurred the 'space race'. Sputnik 1, the world's first artificial satellite, the size of a basketball and weighing only 183 pounds (83.18 kg), took about 98 minutes to orbit the Earth on an elliptical path. Subsequently, between 1958 and 1960 the US army, for its part, placed four earth satellites into orbit; launched the Free World's first lunar probe and first solar satellite; launched three primates into space, two of which were recovered alive; initiated effort on a 1.5-million-pound (c. 6,818 kg)-thrust booster being designed for a lunar exploration vehicle; and began work on the launch vehicle which would carry the first men into space.

Stage II: Mail delivery

Aviation industry

Since its use of airmail contracts during the First World War, the US Post Office Department has long been praised for promoting aviation and its respective technologies. In 1918, the US Post Office Department took over the airmail service from the USAAS (US Army Air Service). Initially, the Department used surplus war planes, which were for the most part flimsy and not built for long, cross-country flights. Consequently, for the purposes of airmail, planes with a larger operational range and payload capacity were introduced, such as the Airco and the Junkers JL-6. Better planes were not the only improvements in the aviation industry to come out of the US Post Office Department, however: in 1921, 10 radio stations were installed along the New York–San Francisco route to transmit weather forecasts, and by 1924 flights were guided by a transcontinental airway with rotating beacons and illuminated emergency landing fields along the way. NASA chief administrator Michael Griffin has been known to compare the forecast role of NASA in the commercialization of space to the role played by the US Post Office Department in the advancement of aviation. Despite this comparison, Griffin has expressed concern that the current stability of the state market may be the fork in the road where the similarities of the two industries part: 'What we have not had is a stable, predictable government market for space services sufficient to stimulate the development of a commercial space industry analogous to that which was seen in the growth of aviation' (Pritchard and MacPherson 2005).

Space industry

In 1994, the Commercial Space Transport Study was released by a group of leading aerospace contractors. The study documented the need for faster package transportation globally in today's worldwide market and included interviews with Federal Express and many other courier services. Several of these companies expressed confidence that there would certainly be a market for the fast package delivery that a sub-orbital rocket-plane could offer (Zuprin 1998). A rocket-plane, made possible by the availability of reusable rocket engines, has the capacity to take off and land from conventional airports, but flies out of the atmosphere at a supersonic speed, and zooms through sub-orbital space before re-entering the Earth's atmosphere and landing as a conventional plane at an airport (Zuprin 1998). These space planes have already made package delivery possible; In August 2007 an unmanned Russian cargo ship carried more than 2.5 metric tons of supplies to the International Space Station, including books, movies and gifts for the crew.

Commercialization of both industries

Now that we have discussed briefly the similarities of the aviation and space industries, we shall spend a little more time analyzing the commercialization of the aviation industry in an effort to understand what may be likely to happen within the space industry in terms of overall strategy and commercial structure.

The commercialization of the aviation industry

In 1926, the Air Commerce Act was introduced, which was the foundation of the state regulation of civil aviation in the USA. The Act came about as a result of pressure from the aviation industry, which believed that air transport could not reach its full potential without government intervention to improve and maintain safety standards. The Air Commerce Act encouraged the formation of new commercial airlines, including:

- Northwest (1926);
- Eastern (1927);
- Pan American (Pan Am) (1927);
- Boeing Air Transport (1927) – subsequently became United (1931);
- Delta (1928);
- American (1930); and
- TWA (1930).

This commercialization was facilitated by the emergence of automobile manufacturer Ford's all-metal Tri-Motor Aircraft – the 'Tin Goose'. The 'Tin Goose' could fly at 110 mph (177 kmph) and hold 12 passengers. Between 1926 and 1935, flying boats (seaplanes) opened new markets for air-travel. Pan Am, with the use of these craft, cleared the way for the first scheduled trans-ocean travel. The first trans-Pacific flight took place in 1935 and the first trans-Atlantic flight in 1938. Commercial aircraft of the 1940s were derivatives of military transport planes and bombers. Subsequently, the 1950s saw the development of the turboprop engine, which led to the first propeller-driven aircraft. Shortly after this, the turbojet engine emerged, which consequently led to the large commercial jet aircraft of the 1960s that are still widely in use today. Subsequent to the deregulation of many services sectors, the 1990s saw a dramatic increase in the number of strategic alliances between airline companies, a pattern we believe may repeat itself in the space industry. Before putting forward our projections for the space industry, we shall recapitulate briefly on the history of some of the major strategic alliances that have shaped the airline industry into the form it has today. Throughout the 1990s, national governments reduced their control over route allocation and pricing. As a result, many of the large air carriers chose to enter into agreements with competitors in an effort to reap the benefits of cooperation. This allowed them to extend their international reach by facilitating a wider mass and global presence, thus creating value through knowledge and quickly learning about unfamiliar markets (Hertrich and Mayrhofer 2005). Despite the new freedom that came with deregulation, there were certain barriers to entry which resulted in the formation of alliances rather than mergers or acquisitions. Some of these barriers included the fact that established major air carriers, more specifically those from the USA, typically dominated the market because of the importance of their own domestic market. Also, certain government bilateral agreements limited the ability of several airlines to serve a number of markets. Finally, prohibitive antitrust rules and strong corporate cultures often acted as barriers to mergers and acquisitions (Hertrich and Mayrhofer 2005). As a result, most airlines opted for strategic alliances that allowed them to gain access to new markets without heavy investment, and so the 1990s saw a series of interlining (interline ticketing) and code-sharing activities (code-sharing is the process of one airline selling seats provided by another airline) (Hertrich and Mayrhofer 2005). Below we can see some of the alliances that were formed during the 1990s:

- 1994: 'European Quality Alliance'/(Qualifier) consisting of Austrian airlines, SAS and Swissair;
- 1997: 'Star Alliance' consisting of Air Canada, United Airlines, Lufthansa, SAS (who left the 'European Quality Alliance') and Thai Airways.
- 1999: 'One World Alliance' consisting of AA, BA, Cathay Pacific, Canadian Airlines and Qantas.
- 2000: 'Sky Team' consisting of Aero Mexico, Air France, Alitalia, CSA Czech Airlines, Delta Airlines and Korean Air.

All these exceeded their bilateral interlining and code-sharing agreements, engaging in a number of supplementary activities including using their membership as a marketing and communications tool.

The commercialization of the space industry

As we have discovered, the commercialization of space has only come about after a long period during which the industry was reserved solely for the purposes of national interest. Space-related organizations were owned and run traditionally for reasons of national prestige, thus bearing a striking similarity to the aviation industry. As a result of the similarities that have been identified between the two industries, we feel it likely that the commercial space industry will see some major consolidation over the next few years. In an effort to compete, many of the smaller companies may join forces and it may be that only a few large and dominant groups, perhaps specializing in different competencies (that is, space tourism, space internet, extracting precious metals and so on), will remain.

There has already been collaboration on a small scale, particularly on the part of NASA, as we saw previously. However, NASA has already been in contact with the Virgin Group in an effort to buy flights for microgravity experiments (Morring 2006). We feel that it is this type of partnership that is bound to surface initially, and it will also characterize the first stages of collaboration in the commercial space industry. It is likely that during this early stage many partnerships will be formed in an attempt to leverage technological capabilities as opposed to quickly learning about unfamiliar markets, which was one of the primary reasons for the strategic alliances that took place within the airline industry. As we have already noted, there exists potential within the space tourism industry for an oligopoly; however, if this was ever to become a likely event, one would imagine that relevant regulatory bodies would intervene to prevent such an outcome. If that was the case, then consolidation would provide a viable means for companies to increase their market share. There are many benefits to having a significant market share in the space tourism industry, specifically for the cost advantages that arise from such a market position and the opportunities available to reinforce existing barriers to entry that arise. This is partially because the aviation industry only evolved commercially after the best part of a century, while the commercial space industry has come about much faster and so the technology is still relatively young. Furthermore, the importance of technology in the space industry is such that any opportunity to leverage technological capabilities or develop new competences would be much more attractive to companies within that industry rather than an opportunity to compete for a different group of customers. Also, the different cultures involved in the aviation industry are far more diverse, as virtually every country in the world has at least one airport, if not an airline. The space industry, however, seems to be dispersed more by region. The EU, NAFTA and the East Asia EPA are the regions with the highest concentration of space-related commercial activity, thus minimizing the number of cultures involved. Furthermore, the general offering is undoubtedly going to be highly standardized, given the inherently risky nature of the activity, which further suppresses issues of cultural arbitrage.

Convergence between the two industries

There is a possibility of convergence between the airline industry and the burgeoning space tourism industry. The underlying forces behind this potential convergence are twofold. First, supply-led convergence could result from changes in the external business environment of firms in either industry, specifically in the areas of regulation and deregulation. Supply-led convergence occurs where organizations begin to behave as though there are linkages between the separate industries (Johnson *et al.* 2008). A positive change in the level of deregulation in the airline industry could lead to a wave of alliances and consolidation, which could be repeated in the space tourism industry given its current low level of regulation. The space tourism industry is still in its infancy, with real competition yet to get fully under way, but the favorable legislation laws in place could attract the attention of experienced players in the airline industry. The possibility of an oligopoly in the space tourism industry is a very real one if the industry continues to adhere to stable market conditions.

Second, demand-led forces could lead to the convergence of the airline and space tourism industries. When customers begin to behave as though industries have converged, such convergence is said to have been led by demand forces (Johnson *et al.* 2008). The key element of demand-led convergence between the space tourism and airline industries will center on the substitution of one product for another, in this case sub-orbital for airline flights. This is a distinct possibility, in particular for a company such as the Virgin Group, which has a presence in both the airline and space tourism industries. The convergence of these separate industries could be brought about in its initial stages by the use of chartered flights using spacecraft for point-to-point travel around the planet, an issue we shall address later in this section. Certainly, it seems that Virgin Galactic, the major player in the sub-orbital space tourism industry, has an interest in such an outcome.

The civil aircraft industry needs high global integration and responsiveness, and low local responsiveness. This form of inter-national strategy is associated with a global orientation. A global strategy most commonly involves a high degree of standardization, with an emphasis on centralizing decisions at headquarters and creating value by designing an offering for a world market while maximizing economies of scale through effective and efficient exe-cution of manufacturing and marketing activities.

As discussed in relation to Virgin Galactic, minimizing costs is certainly an objective for those competing in the commer-cial space industry and, as global strategy dictates, there will probably be a large amount of standardization involved in the offering. This is most likely to be the case for Virgin Galactic's competitors, who, in attempting to catch up with the market share captured by Virgin as a result of its first-mover advan-tage, and because of the emerging nature of the space tourism industry, will be pressurized into imitating the market leader. This will lead to homogeneity and standardization throughout the sub-orbital space tourism industry (Porter 1996). As such, it is probable that the commercial space tourism industry will also adopt a global strategy as it is not only the strategy adopted by civil aviation but it also corresponds to the type of operation best suited to the commercial space industry.

The future of the commercial space industry

Our primary reason for examining the link between aviation and space as analogous industries was in part a result of our final projection regarding the next step for commercial space. Aviation evolved from military aircraft to postal aircraft to its final stage of a global passenger service. We feel that commercialized space travel will probably reach the same conclusion, but in doing so will render traditional aviation obsolete. The reason behind this shift from the aviation industry to the commercial space industry lies primarily in the cost and scarcity of jet fuel. Dr Campbell, the former chief geologist and vice-president of a string of major oil companies including BP, Shell, Fina, Exxon and ChevronTexaco, claimed in 2005 that the peak of regular oil 'the cheap and easy to extract stuff' had already come and gone. He estimated that even when one factors in the more difficult-to-extract heavy oil, deep sea reserves, polar regions and liquid taken from gas, the crisis will come as soon as 2011 (Howden 2007). As a result, fuel surcharges in conventional sub-sonic aircrafts are ever-increasing. On 7 November 2007, United Airlines released a statement revealing that they would be introducing a US$5 one way and US$10 dollar round-trip surcharge because of rising fuel costs. The airline explained that for every dollar increase in the price of crude oil, annual costs for the company go up by US$65 million. With crude oil prices soaring by 42 percent since August 2007, the continued use of jet fuel in conventional aircraft is simply unsustainable (Raine 2007).

In contrast, liquid oxygen, which is extracted from the oxygen found in the air through a process of fractional distillation, constitutes most of the rocket-planes' propellant fuel, and costs 10 cents per kg – a quarter of the price of the jet fuel that a conventional aircraft consumes. Some estimates reveal that the cost of a ride on a sub-orbital rocket-plane to travel from one country to another is likely to be less than double that of a traditional sub-sonic aircraft ride (Zuprin 1998). The premium paid by the customer would cover the benefit of such conveniently rapid travel – which, thanks to the supersonic speeds made possible by sub-orbital voyaging, would have the capacity to offer a flight from New York to Paris in under one hour – as well as the experience of the sub-orbital flight itself, which would include a view of black, starry space and the feeling of weightlessness as the craft enters the Earth's orbit on descent (Zuprin 1998). Consequently, one could argue that the merging of aviation and space could come about not only as a result of increasing wealth leading to an increasing demand for rapid travel, but also out of a functional need brought about by the obsolescence of aviation. Alex Tai, Virgin Galactic's Chief Operating Officer, has already said that the commercial sub-orbital space flight operation the company is currently putting together is to be a stepping-stone to the next stage of space activity they wish to explore – point-to-point travel around the Earth (Roach 2007). There have already been talks between Virgin and NASA regarding the feasibility of a hypersonic passenger service, and both parties are exploring options regarding a collaboration, whereby Virgin would pay for the project and NASA would take care of the requisite research (Roach 2007). 'The long-term goal – going from A to B – is probably where the larger market is,' states Tai, who appears confident that a global passenger service is going to be the big money-spinner in the future of the commercial space industry. Because of the arguments we have put forward, as well as the enthusiasm of some of the biggest players in the industry, we are inclined to agree.

Forces affecting the industry

To better understand the space industry, Porter's Five Forces Framework Model (Porter 1996) will be used. Porter's Five Forces focus mainly on industry structure analysis in an organization's external environment. It reveals the source of competition in the industry and external influences, including the threats to the industry and opportunities to obtain competitive advantage. In the dynamic and competitive space environment, survival, growth and profitability are of the essence for all involved. This section presents an overview of Porter's Five Forces Model, with an examination of the space industry as a whole.

Existing competitive rivalry between suppliers

There has been significant advancements in civil space programs, which have fostered both international cooperation and technical and scientific achievement, but also driven geostrategic competition. In recent years, changes in funding and policy priorities of several space programs indicate the growing rivalry in space, particularly in human space flight and lunar exploration. In 2003, China became the third country to launch a human being into space, and India has since proposed a human spaceflight program. The USA, Russia, Japan, India, China and the European Space Agency have each announced plans for future lunar exploration. Whether these announcements will bear fruit, or if the new space race is real or imagined, the military tensions that drove the first space race cannot be ignored. Cooperation and rivalry in space tend to follow geopolitical patterns on Earth, and there are indications that strategic partnerships are strengthening. Of note is the relaxation of US trade restrictions on sensitive space technologies to India at the same time that China is working with key allies such as Pakistan, Nigeria and Venezuela. There is an aim to reduce the potential for confrontation in space, but as the number of players increases and the stakes get higher, it becomes more difficult to manage political and military tensions (Foust 2003). The EU and Japan have the economic and technological means to deploy weapons in space, but they lack both the political will to challenge the USA and the ability to fund the costs of an independent defense policy. That said, the ESDP (European Security and Defence Policy) and the satellite plan Galileo (the alternative to the American Global Positioning Satellite, GPS) have provoked irritation in their main ally (Garibaldi 2004). Russia has the know-how to compete militarily in space but lacks the financial resources. If it had the means, Russia would presumably put into effect a space policy aimed at filling the power gap with the USA and attempt to reestablish a multi-polar international order.

Threat of new market entrants

The number of countries involved in space exploration has grown from a small, select group in the 1950s to more than 80 nations that today have organized efforts to use space exploration to benefit their own people. The future of space exploration will be based on such international involvement and, more importantly, on collaboration among nations to benefit people everywhere. Nations that are capable of increasing competition in space exploration are notably Japan, China and India. Despite China's funding not being in the same league as the ESA or NASA, the successful manned space flights of Shenzhou 5 and Shenzhou 6, and plans for a space station as part of the Chinese space program of the People's Republic of China have shown what the country can accomplish. The US military is evidently keeping a close watch on China's space aspirations. China's 2007 ballistic missile launch to destroy a satellite was taken by some observers as a token that the space race had never really ended and in fact had only expanded.

The bargaining power of buyers

Thus far, governments have played a primary role in the space industry, therefore the bargaining power of buyers has been low. Governments are the leading advocates and financers of space discovery, especially in the BRIC nations; the ESA member states all contribute according to set budget specifications; and NASA's bureaucratic desire to monopolize space is well-documented. However, following harsh criticism over huge budgets and red tape fiascos, there have been hints regarding the privatization of NASA in the near future, as the private sector may offer lower costs and greater innovation than the government. The USA, in the light of recent continuing developments and innovations by China and Russia, can ill-afford to remain passive in an economy dependent on satellites and aerospace. Therefore, if NASA is no longer at the cutting edge, it should be turned over to those who are, namely the private sector. An Asian space race is providing new competition to NASA, reigniting a spark that was once considered moribund. Perhaps this is the new lease of life that NASA needs to become innovative once again. The USA had planned a manned lunar mission by 2025, but there are other countries with designs on the moon. After two successful manned missions into space in 2005, the moon is in China's sights. Japan, India and the USA's old rival, Russia, all have active space programs, with national pride, national security and even commercial gain all at stake. 'There's a mini-space race going on in Asia with Japan, China and even India claiming an interest in sending astronauts to the moon,' Bill

Read of the Royal Aeronautical Society told CNN. So, while private interest is continually growing in the space arena, governments' predominant position of control has meant that bargaining power remains relatively low for buyers.

Power of suppliers

The power of suppliers in the space industry is high – especially in the highly developed nations. NASA's technological capabilities are unmatched by any other space player, but its source of competitive edge may be slipping, as emerging players such as China and India are developing their technological departments at an alarmingly fast rate. The fact that a wide range of technological transfers are outsourced to India has allowed for learning and vast improvements in their own technological capabilities. Their development of a probe to scan the moon's surface in greater detail than ever before has resulted in a great amount of interest from all key space players. Russia's outstanding experience and capability in the space sector, allied to its recent economic resurgence, has meant the country is technologically sound and capable of giving the USA a run for its money in space development, and reducing the Americans' supplying power significantly. China in particular will emerge as a major space player, mastering the full range of space technologies, and is likely to generate the world's largest demand for space infrastructure. There is money to be made in space. According to a report by the Centre for Strategic and International Studies released in 2010, the 'space economy' is estimated to be worth about US$180 billion, with more than 60 percent of space-related economic activity coming from commercial goods and services. 'Space has always been commercial. Two-thirds of the satellites today are commercial so big money has been made from space technology. Space tourism is a new part of space's business sector that might be small now, but it will grow' (CSIS 2010, p. 47).

Threat of substitutes

The threat of substitutes will affect different parts of the industry in varying ways. Solar power energy is of huge interest to countries wishing to harness energy and distance themselves from the recent oil crisis that has hit the Western economies. Solar power is just one of the energy resources that can be chosen to fulfill energy requirements on Earth. Other, perhaps more appealing, options include wind power, geothermal, gas or oil – all of which are more readily accessible on Earth. In relation to the space tourism industry, there is a colossal threat of substitution. It is stating the obvious, but not everyone's dream vacation is to orbit the Earth, and paying US$20 million for the trip. Space tourism is conspicuous consumption at its highest! As of 2010, space tourism opportunities are limited and expensive, with only the Russian Space Agency providing transport. Virgin Galactic is one of the leading potential space tourism groups today. Galactic will be the first private space tourism company to send civilians into space on a regular basis. There was considerable interest in the 2009 space adventures – however, space tourism will not make Branson a sizeable fortune in the forseeable future.

Key strategic issues facing the industry: PESTEL analysis

To gain a slightly deeper insight into the future of the space industry we adopted a PESTEL – Political, Economic, Social, Technological, Environmental and Legal – analysis.

Political

Traditionally, the space industry is a government area of spending. This has been changing over recent years, but governments across the world have been investing heavily in the industry. This could be because governments see a successful future for themselves within the industry of space. However, another factor we did consider is, with all this government interest and spending is there a potential for a battle of the super-powers within the industry? As noted earlier, there are many large and powerful countries and states involved: the USA, Russia, Europe, China, Japan and India. They are all competing within the one industry, and the potential for major competition is a factor to be considered.

Economic

With regard to the space industry, it is the stronger economies that have more money to invest and this in turn leads to them making more discoveries. With these discoveries, their economies become even stronger. Most governments seem to be very aware of this trend. Emerging economies such as China and India are using their economic growth to break into the space industry and make technological advances. They too seem to be aware that this industry has a huge potential.

Social

Cultural change

There are increasing numbers of people from across the world getting involved in the space industry and want to be involved in the industry's growth. This could be an indicator that the space industry might become a uniting front for the world as there is no similar precedent. The industry has an effect on people the world over.

Expectations

It seems that more people want to be involved because of their expectations regarding the idea of space tourism. People seem to believe at this point that there will be a possibility for all to take part in this aspect of the industry.

Demographics

The demographics of the space industry have been changing over the years. Increasing numbers of people without a major technological background are becoming involved, particularly with regard to space tourism. However, at the time of writing this option is only available to the super-rich who have sufficient money to invest. This could be the cause of more social division and social stratification in the short term.

Technological

Within this industry there are continuous technological advances. Without these, the growth of the industry would grind to a halt. There are, however, major safety concerns surrounding these advances. There is also a call for more reusable space products; for example, with regard to launch testing. This in turn would be more economic. The transport from Earth into orbit requires huge investments in technology, training and financial resources. In the majority of cases these technologies are highly specific as space technology is enormously complex, therefore the development of all new technology is extremely costly and time-consuming. Each step forward in the industry can take many months, even years, to come to fruition, and accidents such as that of the Space Shuttle Columbia, which halted the operation of NASA's Space Shuttle for months, can be detrimental to progress. If the future of the space industry is in fact space tourism, rapid developments in technology need to take place. At present, human space flight is still considered to be relatively dangerous; much of this risk needs to be eliminated before the first human tourists are taken into space. Simply getting the public into space safely is the challenge in the immediate future as far as technological advancements in the industry are concerned. And the idea of hotels, leisure facilities and tourist resorts raises other technological difficulties. Thus the development of the space tourism industry is going to be a long and costly process. The highly specific complex technology is, however, necessary for the industry to succeed. Since safety is the number one priority in this case, it is unlikely that the technological advances will be rushed. There is also the issue of financing all these developments.

Environmental

It is highly likely that many environmental issues will arise in relation to space travel. There is little information currently available on the effects of space exploration on the environment, but it is unlikely that such effects are positive. In 2003, NASA developed and tested an environmentally friendly rocket fuel that may increase operational safety and reduce costs over current solid fuels. It is unclear, however, if this fuel is now used for all rockets launched, or if it was sidelined. In the future of the space industry a much greater amount of fuel will be used so it would be in the interests of both the space industry and the environment if an environmentally friendly fuel was produced and used. Also, the fact that many of the spacecraft parts are largely used only once and are generally unable to be recycled will raise environmental issues. Focusing on manufacturing products that could be used numerous times or that are recyclable would at least be a huge step forward in reducing the carbon footprint of the industry. This would also provide a huge competitive advantage for the firm that developed these products first. As we saw earlier, in recent years in the airline industry environmental issues have become increasingly important, and customers will now be more aware of the environmental impact of going into space. To combat any negative attention with regard to environmental safety, it is necessary that the industry develops products as soon as possible that are friendly to the environments of both earth and space.

Legal

Legislation concerning space activities is a hazy area. International competition laws are a particular area that many countries have not yet even considered. As the industry continues to grow and develop, the lack of competition laws will give rise to international disagreements. Intellectual property rights are another area of concern for this new area of business. In the space industry, the leakage of knowledge and skills, and the poaching of professionals, may pose a problem. Intellectual property rights help to regulate the area and stop the illegal flow of resources. There is also the issue of safety legislation, which thus far has not been addressed thoroughly enough for this rapidly-growing industry. Security standards as well as compensation for injury or death need to be clarified. Penalties or litigation could do enormous harm to the companies within the industry. The stakeholders of a space company have to bear that in mind when deciding whether to invest in a space venture. As we can see from this, there is little or no legislation governing space. As more and more nations venture into the realm of space the issue of legal ownership is going to arise. As on Earth, where different states have different political and social regulations, the institutionalized world is going to demand some kind of legal structure in relation to space. If tourism becomes as prominent in the future of space as we believe it will, there will have to be a legal framework in place. One of the solutions could perhaps be to assemble an international committee that sets out this framework. If this international regulating board did emerge it would indicate that the society of space was truly globalized.

Further potential: SWOT analysis

The effects of globalization are indeed bringing about changes in the industry. We undertook a brief SWOT analysis to gain an insight into current strengths and weaknesses, and determine their potential for the future as opportunities or threats.

Strengths

The governmental agencies traditionally involved in the industry have built a solid infrastructure. They have provided much-needed resources, funding, knowledge and experience. These factors have all contributed to the social approval and success of the space industry.

Weaknesses

There are currently high barriers to entry within the industry. This is evident following our Porter evaluation of the industry structure. Legislation surrounding the industry is very hazy. Safety issues are a major concern, as are leakages of knowledge and protection of the environment. There is the perception of a high level of risk around the activities of the industry.

Opportunities

As state involvement is weakening, it is providing opportunities for private firms to become involved in the space industry.

The potential for high returns on investment, low regulation by governments, many potential customers, and new market areas for growth and expansion may well confirm what Peter Diamandis, founder of the Ansari X PRIZE, prophesied: 'the first trillionaires are going to be made in space' (Diamandis, 2010). There is now an opportunity to eliminate previous gaps in legislation and address areas such as safety, intellectual property rights and environmental issues. The introduction of private firms to the industry could bring about much-needed changes more rapidly, such as the introduction of cheaper and reusable materials, decreasing levels of uncertainty, and higher efficiency rates. There are also more opportunities for cooperation and collaboration between governmental and non-governmental agencies.

Threats

With increased competition, however, there is a threat of potential hostility among the major state players, turning the industry into a battle of the super-powers, perhaps. There are also increasing danger levels as attempts are made to make bigger and faster breakthroughs.

International business perspective

In forecasting the future directions of the space industry, we have identified three different likely scenarios, each of which is related to a different perspective in international business today:

- The first scenario foresees the final success of globalization. International cooperation among countries is complete, and triggers the inexorable integration of markets, economies and technologies, at least as far as the space sector is concerned. This is the best environment, in which the space industry can develop and become profitable. Space businesses will be able to design a global strategy, as they can provide products and services worldwide without trade restrictions.
- The second scenario sees how regionalism prevails, as in the old era of the Cold War. The world is split into two blocs, and the space industry cannot grow as fast as with international

multilateralism. It is hard for space business to become profitable because of the distorting effect of barriers to trade among countries, based on differing tariffs, rules-of-origin requirements and so on.

- The third scenario shows the collapse of the globalization process, and international relationships become a continuous competition among countries with no cooperation. This is the worst possible business environment for the space sector, which then cannot develop fully its commercial possibilities.

A space firm should engage in international business because of the complexity a space-based economic activity requires. A space firm only operating at a national level is likely to fail, because it is not able to take advantage of the benefits of being international:

- To expand sales: space-based products and services can be standardized, as technological components don't need to fit national tastes – they are not cars, clothes or food.
- To acquire resources: the amazing technological knowledge a space firm requires can best be acquired at an international level, searching for components, technologies, information anywhere they might be.
- To minimize risk: diversify suppliers across countries, or counter competitors' advantages.

In our view, a successful space firm should follow a global strategy. It should expand into foreign operations that champion worldwide consistency, standardization and cost competitiveness. Value is created by designing products for a world market, and manufacturing and marketing them as effectively as possible.

With regard to the space industry, R&D, innovation, funding and risk-taking were the key success factors identified.

R&D is work directed on a large scale toward the discovery of new knowledge around products, processes and services. Before the first satellite and Space Shuttle in the twentieth century, space was a largely unexplored and mostly undiscovered part of the universe. Scientists were largely unaware of the chemical make-up of the elements in space, and space exploration was an alien concept. It was through intense R&D

that the impossible became possible. In the space industry, a large amount of resources are spent on researching new types of materials for satellites and spacecraft, and on the developing of new processes and products. Constant R&D is what is needed if this young industry is to flourish and if its future in tourism is to be realized.

Innovation is defined as the alteration of what is established by the introduction of new elements or forms. It can be thought of as a follow-on to the R&D process. In the space industry it is probably the key success factor. Since the industry is very young and is constantly evolving, innovation is what keeps it from stagnating. Constant invention, experimentation and the creation of new products are what drives, and will continue to drive, the space industry forward. Innovation allows the space industry to improve its methods, to become safer, more successful and ultimately more accessible. It is through innovation that the first satellite and the first spaceship were created, and through continuing innovation that there will be further developments in spacecraft. It is thought that, by creating spacecraft that have reusable (or at least recyclable) components, that the space industry will be revolutionized. Innovation is what will make this creation possible. Innovation is what gives firms in the industry competitive advantage over each other. The most innovative firm is the most successful one.

As for funding, the space industry is extremely expensive to finance. Currently it is only really possible for state bodies or extremely wealthy private individuals to finance firms in the industry. R&D, new product development, raw materials and human capital are just some of the expenses that need to be met on a daily basis. The fact that all the components for the products are so costly and can only be used only once means that financially there is a huge barrier to entry for people thinking of joining the space race. The first fully reusable launch vehicle will cut the cost of a launch by 90 percent immediately and by more as operating and manufacturing experience accumulate. But only through continued funding can this reusable product be developed. It is because of the continued financial backing from governments and private individuals alike that the industry has been able to thrive. As we have seen, funding is a key success factor of the industry but it is also a key strategic issue. Slowly but steadily government funding for the industry is slowing down, and while this leaves a gap for private firms and individuals to participate in the industry, it also raises problems. Some countries that rely heavily on state funding will have to start cutting back on other areas of development, and these countries may not have private parties that are interested in entering the space industry. The high barriers to entry are a deterrent to the non-state bodies who are interested. To achieve successful entry into space tourism the industry is going to need to attract a considerable amount of investment. R&D, materials and technological innovation are just some of the processes that will require a great deal of capital investment in the near future. Certain developments in methodologies and products are likely to create significant cuts in costs for the industry; however, these developments have not yet taken place. The decrease in state funding that is likely in the Western world will be a blow to the industry, but the increased interest from private parties will, it is hoped, help to curb the negative consequences of any such decrease. The space industry must now try to attract more investment from private sources rather than continue to rely on state support that is likely to dwindle.

All industries, especially developing industries, have a degree of risk involved. However, everything that makes the space industry successful is about taking risks. Safety is not guaranteed in human spaceflight. In fact, the process is still considered to be exceptionally risky, and even the most highly qualified and experienced individuals can experience difficulties when on a spacecraft. This is illustrated by the demise of the Challenger and Columbia Space Shuttles in 1986 and 2003, respectively. Without the willingness of scientists to experiment and of astronauts to explore the unknown it is highly unlikely that the space industry could have been as successful as it is today. There is also the financial risk; the space industry is young and relatively unstable, and investment in the area is not guaranteed any return. However, as we saw above, funding is a vital factor for this industry's success, so without the willingness of governments and some private individuals to take the risk of investing, the industry would have been unable to grow and flourish. Even the development of new products and new markets has a degree of risk. There is no precedent set in this industry, since space exploration is such a new phenomenon. It is a risk to assume that the idea of space tourism will be a successful one and to invest so many resources into making it a viable option, but without taking these risks there would be no growth in the industry and no possible creation of new markets. Risk-taking has been and will continue to be the basis for realizing the space industry's potential.

Amin, A. (2002) Spatialities of globalisation, *Environment and Planning A*, vol. 34, no. 3, pp. 385–400.

Barker, C. (1999) *Television, Globalization and Cultural Identities* (Buckingham: Open University Press).

BBC (2009) BBC Science & Nature: – Space, *Exploration Timeline*. Available at: http://www.bbc.co.uk/science/space/exploration/missiontimeline/.

BBC (2007) BBC News Science/Nature: *"Unanimous backing" for Galileo*. Available at: http://news.bbc.co.uk/1/hi/sci/tech/7120041.stm.

Bennett, W. L., Pickard, V. W. and Iozzi, D. P. (2004) Managing the public sphere: journalistic construction of the great globalization debate, *Journal of Communication*, vol. 54, no. 3, pp. 437–55.

Billings, L. (2006) How shall we live in space? Culture, law and ethics in spacefaring society, *Space Policy*, vol. 22, no. 4, pp. 249–55.

Branson, S. R. (2006) *Screw It, Let's Do It: Lessons in Life* (London: Virgin Books).

Brenner, N. (1999a) Globalisation as reterritorialisation: the re-scaling of urban governance in the European Union, *Urban Studies*, vol. 36, no. 3, p. 431.

Brenner, N. (1999b) Beyond state-centrism? Space, territoriality, and geographical scale in globalization studies, *Theory and Society*, vol. 28, no. 1, pp. 39–78.

Broad, W. J. (2007) From the start, the space race was an arms race, *The New York Times*. Available at: http://www.nytimes.com/2007/09/25/science/space/25mili.html?ref=space; Accessed 20 July 2010.

Brocklebank, D., Spiller, J. and Tapsell, T. (2000) Institutional aspects of a global navigation satellite system, *The Journal of Navigation*, vol. 53, no. 2, pp. 261–71.

Burrows, W. E. (1999) *This New Ocean: The Story of the First Space Age* (New York: Modern Library).

Cammaerts, B. (2005) *ICT-usage among transnational social movements in the networked society – to organise, to mobilise and to debate*. Available at: http://eprints.lse.ac.uk/3278/.

Cantrell, J. (2008) Obama administration impact on space policy. *Strategic Space Development*. Available at: www. strategicspace.net. Castells, M. (1996) *The Rise of the Network Society* (New York: Blackwell).

Castles, S. and Davidson, A. (2000) *Citizenship and Migration: Globalization and the Politics of Belonging* (New York: Routledge).

Cerny, P. G. (1995) Globalization and other stories: the search for a new paradigm for international relations, *International Journal.*, vol. 51, p. 617.

CNNMoney (2009) *Business, Financial, Personal Finance News*, CNNMoney. com. Available at: http://money.cnn.com/.

Collins, P. (2002) Meeting the needs of the new millennium: passenger space travel and world economic growth, *Space Policy*, vol. 18, no. 3, pp. 183–97.

CSIS (Centre for Strategic International Studies) (2010) *National Security and the Commercial Space Sector Report*. Available at: http://csis.org/files/publication/100726_Berteau_CommcialSpace_WEB.pdf.

Cunningham, J. (2007) Holidays in orbit, *Professional Engineering*, vol. 20, no. 13, p. 39.

Dick, S. J. (2007) Assessing the impact of space on society, *Space Policy*, vol. 23, no. 1, pp. 29–32.

Dudley-Flores, M. and Gangale, T. (2007) *The Globalization of Space – The Astrosociological Approach*, AIAA Space 2007 Meeting Papers on disk [CD-ROM], AIAA-2007-6076, Reston, Virginia, USA.

Dudley-Flores, M. and Gangale, T. (2009) *Manufactured on the Moon, Made on Mars – Sustainment from the Earth, beyond Earth*. AIAA Space 2009 Meeting Papers on disk [CD-ROM], AIAA-2009-6428, Pasadena, California, USA. Eurospace (1995) Space: a challenge for Europe, *Space Policy*, vol. 11, no. 4, pp. 227–32.

Eurospace/ASD (2008) *Eurospace – The Association of the European Space Industry – Facts and Figures*. Available at: http://pagesperso-orange.fr/eurospace/fandf.html.

Fairclough, G. (2007) *China's Long March to the Moon*. Available at: http://online.wsj.com/article/SB119308504660267440.html.

Ferguson, M. (1992) The mythology about globalization, *European Journal of Communication*, vol. 7, no. 1, pp. 69–93.

Fisk, L. A. (2008) The impact of space on society: past, present and future, New York: *Space Policy*, vol. 24, no. 4, pp. 175–80.

Florida, R. (1996) Regional creative destruction: production organization, globalization, and the economic transformation of the Midwest, *Economic Geography*, vol. 72, no. 3, pp. 314–34.

Foust, J. (2003) *The Space Review: What Is the 'Space Industry'?* Available at: http://www.thespacereview.com/article/34/1.

Friedman, T. L. (2000) *The Lexus and the Olive Tree*, Reprint (New York: Anchor Books). (Originally published in 1999 by Farrar, Straus & Giroux.)

Friedman, T. L. (2005) *The World Is Flat: A Brief History of the Twenty-first Century* (New York: Farrar, Straus and Giroux).

Fukushima, M. (2008) Legal analysis of the International Space Station (ISS) programme using the concept of 'legalisation', *Space Policy*, vol. 24, no. 1, pp. 33–41.

Futron Corporation (2008). *Futron's 2008 Space Competitiveness Index (SCI)* (White Paper). Date of publication: 25 April. Available at: http://www1.futron.com/resource_center/store/Space_Competitiveness_Index/FSCI-2009.htm.

Futron Corporation (2009) *Resource Center*. Available at: http://www.futron.com/resource_center/resource_center.htm.

Garibaldi, G. (2004) The Chinese threat to American leadership in space, *Security Dialogue*, vol. 35, no. 3, pp. 392–6.

Glisby, M. and Holden, N. (2005) Applying knowledge management concepts to the supply chain: how a Danish firm achieved a remarkable breakthrough in Japan, *The Academy of Management Executive*, vol. 19, no. 2, pp. 85–89.

Global Industry Analysts Inc. (2010) *Global Positioning System Report*. Available at: http://www.strategyr.com/Global_Positioning_Systems_GPS_Market_Report.asp.

Goehlich, R. A. (2005) A Ticket pricing strategy for an oligopolistic space tourism market, *Space Policy*, vol. 21, no. 4, pp. 293–306.

Goodhart, M. (2001) Democracy, globalization, and the problem of the state, *Polity*, vol. 33, no. 4, pp. 527–46.

Hannerz, U. and Featherstone, M. (1990) Global culture: nationalism, globalization and modernity, *Third World Quarterly*, vol. 25, no. 1, pp. 207–30.

Harvey, D. (1991) The urban face of capitalism, in J. F. Hunt (ed.), *Our Changing Cities* (Baltimore, Md.: Johns Hopkins University Press), pp. 50–66.

Hawking, S. (2002) *The Theory of Everything: The Origin and Fate of the Universe* (New York: New Millennium Press).

Hertrich, S and Mayrhofer, U. (2005) Strategic alliances in the global airline industry: from bilateral agreements to integrated networks, in P. Ghauri and P. Cateora (eds), *International Marketing* (London: McGraw-Hill).

Hertzfeld, H. R. (2007) Globalization, commercial space and spacepower in the USA, *Space Policy*, vol. 32, no. 4, November.

Hess, R. K. and Kossack, E. W. (1981) Bribery as an organizational response to conflicting environmental expectations, *Journal of the Academy of Marketing Science*, vol. 9, no. 3, pp. 206–26.

Howden, D. (2007) World oil supplies are set to run out faster than expected, warn scientists, *The Independent*. Available at: http://www.independent.co.uk/news/science/world-oil-supplies-are-set-to-run-out-faster-than-expected-warn-scientists-453068.html; Accessed 20 July 2010.

Hughes, K. (1990) Pioneering efforts in space. Available at: http://www.redstone.army.mil/history/pioneer/welcome.html; accessed 20 July 2010.

Hwang, C. Y. (2006) Space activities in Korea – history, current programs and future plans, *Space Policy*, vol. 22, no. 3, pp. 194–9.

IEA (International Energy Agency) (2008) *World Energy Outlook 2008*. Available at: http://www.worldenergyoutlook.org/2008.asp.

ISRO (2009) *Indian Space Research Organisation*. Available at: http://www.isro.org/.

Israeli Minister of Foreign Affairs (2010) Focus on Israel: Israel in space. Available at: http://www.mfa.gov.il/mfa/mfaarchive/2000_2009/2003/1/focus%20on%20israel-%20israel%20in%20space; accessed 8 July 2010.

JAXA (Japan Aerospace Exploration Agency) (2010) *JAXA Repository*. Available at: http://repository.tksc.jaxa.jp/en.

Johnson, G., Scholes, K. and Whittington, R. (2008) *Exploring Corporate Strategy: Text & Cases*, 8th edn (London: Prentice Hall).

Jones, R. A. (2004) They came in peace for all mankind: popular culture as a reflection of public attitudes to space, *Space Policy*, vol. 20, no. 1, pp. 45–8.

Kass, L. (2006) Iran's space program: the next genie in a bottle?, *The Middle East Review of International Affairs*, vol. 10, no. 3). Available at: http://meria.idc.ac.il/journal/2006/issue3/jv10no3a2.html.

Kennedy Project (2009) Kennedy 2 Lunar Exploration Project – K2LX. Available at: http://www.kennedyproject.com/.

Kobrin, S. J. (1997) The architecture of globalization: state sovereignty in a networked global economy, in J. H. Dunning (ed.), *Governments, Globalization, And International Business* (Oxford: Oxford University Press), pp. 146–71.

Lebeau, A. (2008) Space: the routes of the future, *Space Policy*, vol. 24, no. 1, pp. 42–7.

Lele, A. (2005) Pakistan's space capabilities, *Air Power Journal*, vol. 2, no. 1.

Liao, S. (2005) Will China become a military space superpower?, *Space Policy*, vol. 21, no. 3, pp. 205–12.

Loizou, J. (2006) Turning space tourism into commercial reality, *Space Policy*, vol. 22, no. 4, pp. 289–90.

Luke, T. (1996) Identity, meaning and globalization: detraditionalization in postmodern space–time compression, in P. Heelas, S. Lash and P. Morris (eds) *Detraditionalization: Critical Reflection Authority and Identity* (Oxford: Basil Blackwell).

Marx, K. and Engels, F. ([1848] 2002) *The Communist Manifesto* (London: Penguin).

McLuhan, M. and Powers, B. R. (1992) *The Global Village: Transformations in World Life and Media in the 21st Century*, Communication and Society (New York: Oxford University Press). Diamandis, P. H. (2010) *From Space to Energy: Changing the World. For Good.* Lecture delivered at the Massachusetts Institute of Technology (MIT). Available at: http://mitworld.mit.edu/video/331/.Morring, F., Jr. (2006) Dream teams, *Aviation Week & Space Technology*, vol. 165, no. 17, pp. 22–5.

Morring, F. Jr. and Mathews, N. (2004) Third world rising (India's space program), *Aviation Week & Space Technology*, vol. 161, no. 20, pp. 46–9.

Mustafa, N. (2004) A new race for space: look out for Nigeria, *Time*. Available at: http://www.time.com/time/magazine/article/0,9171,993401,00.html.

NASA (2009) U.S. human spaceflight history. Available at: http://www.jsc.nasa.gov/history/hsf_history.htm.

Nelson, R. R. and Winter, S. G. (1982) *An Evolutionary Theory of Economic Change* (Cambridge, Mass.: Harvard University Press).

OECD (2004) Space 2030: Exploring the Future of Space Applications. Available at: http://www.oecd.org/document/18/0,3343,en_2649_34815_34726866_1_1_1_1,00.html.

OECD (2007) *The Space Economy at a Glance*. Available at: http://www. oecd.org/document/4/0,3343,en_2649_34815_39629508_1_1_1_1,00.html.

O'Neil, D. (2009) Space future – general public space travel and tourism. Available at: http://www.spacefuture.com/archive/general_public_space_ travel_and_tourism.shtml.

Onoda, M. (2008) Satellite observation of greenhouse gases: monitoring the climate change regime, *Space Policy* 24, no. 4 (November), pp. 190–8.

Paikowsky, D. (2007) Israel's space program as a national asset, *Space Policy*, vol. 23, no. 2, pp. 90–6.

Peterson, R. B. (1972) A cross-cultural perspective of supervisory values: reply, *Academy of Management Journal*, vol.15, no. 3, pp. 369–70.

Porter, M. E. (1998) *The Competitive Advantage of Nations* (New York: Free Press).

Pritchard, D. and MacPherson, A. (2005) Boeing's diffusion of commercial aircraft design and manufacturing technology to Japan: surrendering the US aircraft industry for foreign financial support, Canada–United States Trade Center Occasional Paper No. 30.

Radhakrishna, R. (2009) India's space programme looks beyond the moon. Available at: http://www.flightglobal.com/articles/2009/02/02/321891/ indias-space-programme-looks-beyond-the-moon.html.

Raine, G. (2007) United Airlines adds fuel-cost surcharge, *SFGate*. Available at: http://articles.sfgate.com/2007-11-09/business/17269906_1_fuel-costs-crude-oil-united-airlines; accessed 20 July 2010.

Roach, J. (2007) Virgin Galactic, NASA team up to develop space-plane travel. *National Geographic News*. Available at: http://news.nationalgeographic. com/news/2007/03/070320-virgin-space.html; accessed 20 July 2010.

Robertson, R. (1992) *Globalization: Social Theory and Global Culture* (London: Sage).

Rutan, B. (2006) Why space needs you. Available at: http://money.cnn.com/ magazines/business2/business2_archive/2006/03/01/8370591/index.htm.

Sawamura, B. and Radke, K. (1992) Future manned systems advanced avionics study COTS for space, *Proceedings of the IEEE/AIAA 11th Digital Avionics Systems Conference, 1992,* pp. 514–22.

Scholte, J. A. (2000) *Globalization: A Critical Introduction* (Basingstoke: Palgrave Macmillan).

Scholte, J. A. (2005) Premature obituaries: a response to Justin Rosenberg, *International Politics*, vol. 42, no. 3, pp. 390–9.

Shapiro, R. (2009) A new rationale for returning to the Moon? Protecting civilization with a sanctuary, *Space Policy*, vol. 25, no. 1, pp. 1–5.

Silva, D. H. D. (2005) Brazilian participation in the International Space Station (ISS) program: commitment or bargain struck?, *Space Policy*, vol. 21, no. 1, pp. 55–63.

Smith, A. ([1776] 1986) *The Wealth of Nations*. Harmondsworth: Penguin.

Smith, A. (2002) Trans-locals, critical area studies and geography's Others, or why 'development' should not be geography's organizing framework: a response to Potter, *Area*, vol. 34, no. 2, pp. 210–13.

Space Adventures (2009) *Space Adventures*. Available at: http://spaceadventures. com/.

SPACE.com (2009) Astronomy and science news and information, astronomy features, astronomy pictures. Available at: http://www.space. com/scienceastronomy/.

Storper, M. (1995) The resurgence of regional economies, ten years later: the region as a nexus of untraded interdependencies, *European Urban and Regional Studies*, vol. 2, no. 3, p. 191.

Suzuki, K. (2007) Space and modernity: 50 years on, *Space Policy*, vol. 23, no. 3, pp. 144–6.

Tarasenko, M. V. (1996a) Current status of the Russian space programme, *Space Policy*, vol. 12, no. 1, pp. 19–28.

Tarasenko, M. V. (1996b) Evolution of the Soviet space industry, *Acta Astronautica*, vol. 38, no. 4–8, pp. 667–73.

Vallaster, C. (2005) Cultural diversity and its impact on social interactive processes: implications from an empirical study, *International Journal of Cross Cultural Management (CCM)*, vol. 5, no. 2, pp. 139.

Van Rooy, A. (2004) *The Global Legitimacy Game: Civil Society, Globalization, and Protest* (Basingstoke: Palgrave Macmillan).

Vick, C. (2009) North Korea's space, ballistic missile administration, development infrastructure. Available at: http://www.globalsecurity.org/ space/world/dprk/agencies.htm.

Yip, G. S. (2003) *Total Global Strategy II: Updated for the Internet and Service Era* (Upper Saddle River, NJ: Prentice-Hall).

Wallerstein, I. (1980) *The Modern World-System I: Capitalist Agriculture and the Origins of the European World-Economy in the Sixteenth Century* (New York/London: Academic Press).

X PRIZE Foundation (2009) *X PRIZE Foundation*. Available at: http://www. xprize.org/.Zak, A. (2008) Russian space program in the first decade of the 21st century. Available at: http://www.russianspaceweb.com/russia_ 2000_2010.html.

Zhao, Y. (2005) The 2002 Space Cooperation Protocol between China and Brazil: an excellent example of South–South cooperation, *Space Policy*, vol. 21, no. 3, pp. 213–19.

Zuprin, R. (1998) Aviation's next great leap, *MIT Technology Review*, vol. 101, no. 1, pp. 30–6.